Serafina Loves Science!

QUANTUM
QUAGMIRE

D0880456

Serafina Loves Science!

QUANTUM QUAGMIRE

Cara Bartek, Ph.D.

Absolute Love Publishing

Absolute Love Publishing

Quantum Quagmire

A book by Absolute Love Publishing

Published by Absolute Love Publishing
USA

Illustration by Logynn Hailley

ISBN-13: 978-0-9995773-4-9
United States of America

By Cara Bartek

Cosmic Conundrum
Quantum Quagmire

Dedication

To my silly willy, Penny

Praise for Serafina Loves Science!

"Through the quirky character of Serafina, *Cosmic Conundrum* shows us that science can be fascinating! Girls, you can embrace your curiosity, follow your passions, and most importantly, just be you! As an environmental scientist, I love this series!" - *Emily Thompson, Edwards Aquifer Authority Environmental Scientist*

"Every opportunity for inquiring girls to stretch their STEM skills and reach for new horizons is time well spent. The unique characteristics of the exceptional in-between-agers portrayed in Serafina Loves Science! will certainly strike a chord with the readers, and Serafina's adventures are sure to inspire new ideas and open new pathways of creativity." - *Rick Varner, Scobee Education Center Director*

"This book is fabulous! It made me laugh out loud a few times. My niece Sadie said she loves that Sera has GRIT! This is a true example of how we can let bullying define us or inspire us. I also loved that it makes science fun! I cannot wait for the sequel!" - *Rebecca Boenigk, Neutral Posture CEO*

"*Cosmic Conundrum* is an excellent book for all students. It is about Serafina's love of science and her adventures in pursuing her dream. The book is full of science information from the perspective of a young girl and also addresses several issues common to our youth, including friendship, loyalty, facing one's fears, pursuing one's dreams, and dealing with bullies. I highly recommend this book for young readers, especially girls. *Cosmic Conundrum* is an exciting and entertaining story that will give girls confidence to pursue science careers." - *Miguel Sepulveda, school counselor and STEM Coordinator*

"*Cosmic Conundrum* is a book for all ages, a timeless story about humanity and kindness with a technological twist of events. Fascination with science and space exploration leads Serafina, a precocious thinker and oblivious-to-girlie-things 11-year-old, to a space adventure camp. Will she make friends? Will she win the top prize for her invention? Will she be chosen to be a junior astronaut? Mixed emotions surge when the unexpected happens to Serafina." - *Lidya Kushner Osadchey, business consultant, executive coach, parent, and community leader*

CONTENTS

HELLO!

My name is Serafina Sterling, and I love science! I don't just like it. I love it. I love biology and chemistry and entomology and astronomy and ichthyology and herpetology and horticulture and genetics and geology and botany ... DEEP BREATH ... and the list goes on and on and on, much like my favorite number: pi. You *do* know that pi never ends? It doesn't! Well, theoretically it *could*, but there are computers still trying to figure out when that might happen so I guess it's a mystery that can only be solved by ... SCIENCE!

Which is just one more reason why I say I LOVE science. This isn't the same kind of love a person has for ice cream or soccer or the color yellow. This is a love that is strong and lasting and real. This love is as hard and strong as wurtzite boron nitride, a substance that is 58 percent harder than diamonds and was made by scientists in a lab using nanomaterials. That's even harder than my big brother Apollo's head!

You see science helps me understand the world. Because sometimes, okay, lots of times, my world can be confusing. I'm in sixth grade, but sometimes

I'm faced with problems that are beyond the sixth grade—grown-up problems. Like the problem my best friend, Tori Copper, just told me she's having. It's a big, grown-up problem that will change her life forever. And I have no idea how to help her. There aren't many things kids like us can do about grown-up problems. But, we do have one tool at our disposal. This is a story about how I helped my best friend, Tori Copper, make it through her parent's divorce using the thing that I love most: SCIENCE!

CHAPTER ONE

Names can be strange. Sometimes they tell very little about a person. Sometimes they tell a lot. Take, for instance, Albert Einstein.

When someone calls someone else an "Einstein" that usually means the person is brilliant—even if the person saying it is being sarcastic. Like that time I correctly answered a question in class about the speed of light (299,792,458 m/s) and that snooty Todd Brakefield stuck out his tongue at me and said, "Way to suck up, *Einstein*."

Todd didn't mean it as a compliment, but I definitely took it as one. In fact, it's the greatest compliment a budding scientist like me can get because Einstein was a brilliant scientist. However, before Einstein ever made his groundbreaking discoveries about general relativity, he was just good ol' Albert who liked to do math. The name Einstein meant nothing.

Then there are other times when names can tell you a lot about a person. Take my best friend, Tori Copper. Not COOPER, but Copper, like the 29th element on the periodic table. Tori Copper *is* the color of copper. Her hair, her freckles, her lips, even her favorite rain boots are all the same color as copper. Not red and not orange, but a color in between. She even has some of the properties of her metallic namesake. She's strong, slightly conductive, and sometimes the atmosphere can tarnish her a bit.

I should have known something was wrong after Tori's birthday party. I'm no federal agent, but there were clues. Maybe if I had put two and two together earlier, things would be different.

"Serafina, are you going to stop?" Tori asked as I came rolling toward her and a group of our friends.

We were at the local skating rink, Smooth Groove, for Tori's 11th birthday. It was a pretty cool place. The skates came in every shade of neon and everything else was painted in psychedelic hippie colors that glowed in the black lights. The commercial said you could, "Glow as you roll," which proved to be an accurate statement as our green, blue, yellow, and pink skates lit up the wooden rink.

A group of six girls in candy-colored skates were gathered around Tori giggling and dancing. "It's fun to stay at the Y-M-C-A," they sang, shaping the letters with their arms.

"I can't stop!" I screamed. "I can't!"

"Turn, Serafina! Turn!" Tori yelled.

"I can't turn!" I flapped my arms wildly at my sides. "Too much forward momentum!"

"Don't you mean velocity?" Roger Penright said from the snack bar. He held his chilidog and snow cone high in the air and smirked. "No such thing as forward momentum. Only velocity. Try opening a physics book every once in a while, Serafina."

"I don't need this in my life right now, Roger. I'm going to die!" I screamed as I rolled faster than I had ever rolled in my life. "And I'm way better at physics than you'll ever be!"

Roger rolled his eyes and took a bite from his chilidog. "Says the girl who can't stop her skates."

Where are the brakes? I wondered. *Skates must have brakes!*

"Serafina, you're going to kill us all!" Tori shrieked in a panic. Tori seemed to be the only one who had noticed the impending doom. The other girls were still dancing and singing.

"Try changing directions!" Roger yelled as I whizzed past him.

"Try chewing with your mouth closed!" I yelled back.

"Turn, turn!" Tori yelled.

It was too late. "I can't!" I screamed. My speed had reached Mach 1, and the swiftly approaching wall was the only thing that was going to stop me. The only problem with that was that there were six dancing partygoers in poufy skirts between the wall and me. And they still hadn't noticed me.

"It's fun to stay at the Y-M-C-A," sang the girls.

I was going to squish one of my friends and totally flatten those poufy skirts.

"MOVE!" I shouted as I careened toward them on my glowing, blue skates.

They finally noticed me. Arms froze in place and mouths dropped open.

"MOVE!" I yelled again, flailing my arms in a useless attempt at slowing down.

They began to scream and run in opposite directions. Some headed toward the rink while others scaled the wall between the rink and the snack bar.

"She's coming right toward us!" one girl yelled.

"We're all going to die!" another exclaimed.

"No, no, NOOOOOO!" Tori screeched and threw

her hands in the air, but it was too late.

Tori Copper—the person of the hour, the reason why we were at the skating rink, my best friend in the whole wide world—was the only one who was unable to avoid my super-sonic-speed death roll. She threw her arms in the air like she was making the "Y" from the "YMCA" dance. I squeezed my eyes shut as my velocity flattened her into a pancake.

I slowly opened my eyes and saw the tip of Tori's nose right against mine.

"Ohhhhh," she moaned.

I rolled off of my friend and brushed off my knees. I said the only thing that came to mind. "Happy birthday."

Tori rubbed her head and smiled. "You are the worst skater."

I giggled. "I know. But I have a lot of other good qualities."

Tori pushed herself up on her elbows. "Oh yeah? And what are those qualities?"

I furrowed my brow and thought hard. "I'm a snappy dresser."

"Yeah, right," she chuckled. "Keep going."

"Hmmm ..." I rubbed my chin like an old-timey detective from one of those British crime shows on PBS. "I always chew with my mouth closed."

"That's better than Roger," she said.

"Hey," called a chilidog-muffled voice from the table closest to us. "Not cool!"

Tori laughed. "Keep going. What are your other good qualities, because right now I'm lying on my butt at my 11th birthday party and I'm pretty sure I broke something. Something important ... like my pelvis or my frontal lobe."

I clapped my hands together. "I have just the thing." I rose on my shaky legs and grabbed onto the nearby wall. I offered my hand to Tori who refused. I guess she wasn't ready to trust my standing ability on the skates either. "I got you the best birthday present a person could ever ask for. I'll get it now."

I raced, or rather stumbled, toward the gift table, pushing the frowning, poufy-skirt girls out of the way. They still seemed to be peeved from my almost accidental skating homicide. I wasn't sure what they were so upset about. I bet if I was in the right jurisdiction with the right judge, I could have gotten away with a simple misdemeanor.

I frantically scanned the table and found my package. My mom had wrapped the gift for me with shiny, blue paper. She had carefully tied a silver ribbon in an intricate pattern. I admired her handiwork for all of two seconds before I began to rip into the package.

"Sera, I'm supposed to do the unwrapping," Tori moaned from behind me.

I turned to see my equally shaky-legged comrade had finally made it to the gift table.

"I know. I know. But I couldn't wait."

I quickly extracted the tiny paper card and handed it to her.

"What is this?" she asked as she scanned the present.

"Just look." I grinned ear to ear.

Her eyes flicked. "Oh ..."

I began to jump up and down. "I know!"

"You got me a ticket to the SOLAR ECLIPSE?" Tori screeched.

I nodded my head up and down furiously. "Yes!

We're all going as a family and my mom and dad said it was totally okay if you come. We are going to watch it from Yellowstone Park so we can see geysers, and elk, and maybe see if we can discover a Bigfoot or something. And the whole way we can listen to music and watch movies, and I already told Apollo you and I get the whole third seat of the minivan all to ourselves, and if he so much as—"

"I CAN'T BELIEVE IT!" Tori screamed and jumped up and down.

"We're going to have so much fun!" I shrieked. "Here, put on your viewing glasses." I pulled a pair of dark shades from the envelope.

Tori placed them on her face. "I can't see a thing," she said.

"That's okay because you really don't want to burn your corneas."

"Mom! Dad!" Tori yelled as her parents walked up to us. "Serafina is going to take me to see the eclipse in Wyoming for my birthday!"

"What?" asked Tori's mom.

"That's awesome," said Tori's dad. "You're going to have a great time."

"Jason, wait. We can't just let her travel across the country all by herself," Tori's mom said. "This is something we need to discuss as a family."

Mr. Copper dismissed the statement with a flick of his wrist.

"If it will make you feel any better, Miss Copper, Apollo has agreed to give us the third row. And I have already furnished Tori with doctor-approved eye protection so she won't burn her eyeballs—"

"Shhh," Tori hissed. She was making a strange face and it appeared all her freckles were converging in a copper-colored flush. "Not helpful."

Mr. Copper resumed, "I don't see what the big deal is. The Sterlings invited Tori on a nice little road trip and I think—"

"Well I think this is something we should discuss as a family before we just—"

"Do we really have to do this here? At Tori's party?" Mr. Copper asked.

Tori's face grew bright red.

"Do what, Jason?" asked Mrs. Copper. "Be responsible parents?"

"Really? What? I'm irresponsible?" Mr. Copper said in a not-so-nice tone.

"What's going on?" I asked Tori. "I totally screwed up. I'm so sorry!"

"No, no." Tori shook her head in a small, nervous way. "It's not you. This kind of stuff has been happening for a while."

"It has?" I asked.

Tori nodded. "Yeah. But it's no big deal. It usually just blows over. It's really more embarrassing than anything."

"Oh, so they're bickering." I knew all about bickering. My mom said my brother Apollo and I could compete at the Olympic-level in bickering. Bickering *was* no big deal. But ... the nervous way Tori was biting her nails told me another story. Tori only bit her nails before major exams or when she had to get a shot.

It appeared that the Coppers' exchange was attracting a crowd. Not only were the poufy-skirted girls from our school gathered around to watch the fireworks, but people from nearby parties had also stopped to stare.

"I just don't get you sometimes," said Tori's mom.

"Don't get me? Well, I don't get you," Mr. Copper retorted.

Tori continued to chew on her nails. I could tell the fight was really starting to kill the celebration vibe. I had to do something to fix it.

"Come on," I said and grabbed Tori's hand. "You need to teach me how to stop, and I'll teach you how to achieve warp speed."

Tori pulled her nails out of her mouth and smiled. "Okay."

A loud crash rang out near her parents.

"What was that?" Tori exclaimed.

We turned to look. Tori's mom was standing next to the table with frosting all over her hands. Tori's rainbow unicorn cake was on the floor and had a massive split right in the center of it.

"My cake!" Tori cried. "What happened?"

Tori's mother blinked.

"Your mother dropped your cake," said Tori's dad.

"What?" Tori asked in a small voice looking from her dad to her mom and back again.

"Uh, uh," I stammered. This was going to ruin the entire party for Tori, and I just couldn't have that. What with her being my most amazing, best friend in the whole wide world. "Wait! This is totally cool, it's like an existential cake."

Tori's mouth gaped open and she looked at me like I was crazy.

"It says, yeah I'm a pretty unicorn, but I have depth." I scrambled for more. "It's like showing the wounds of modern society. Yeah ... like, you can be the prettiest unicorn out there but also be a mess inside."

"A mess inside?" Tori repeated. "What are you talking about?"

"Philanthropy," I said, shrugging my shoulders.

Tori's mouth continued to gape open.

"Or is it paleontology?" I considered.

Tori closed her mouth. "I think you mean philosophy."

"That's it!" I exclaimed. "Philosophy!"

Tori began to giggle. "You are such a dork, Serafina!"

I smiled. "A dork that's your very best friend," I said and gave her a hug. "Happy birthday, Tori! I hope this year is the best one yet."

CHAPTER TWO

I knew Tori Copper was going to be my forever best friend the first time I saw her holding a bug. Well, technically it was an arachnid. Mr. Hairy Arms, our first-grade teacher, was legendary for his self-aggrandizing lectures. (I learned the word "self-aggrandizing" last month while sitting at the DMV with my mom. Don't ask. It's a long story that involves those little number tickets and their value in modern society.) Mr. Hairy Arms could always be counted on for lots of hand motions, mustache wiggling, and finger pointing. Between the basics of phonics and addition, we were usually subjected to various unsolicited life lessons.

On this particular day, Mr. Hairy Arms was in fine form.

"You see, first graders, ink is just not something you can erase." Wild arm motions. Finger pointing. Mustache wiggling. "You must always use a writing utensil that allows for *redesign* and *reconsideration*." More pointing and mustache twitching.

I rolled my eyes. I felt that I was being singled out during this particular tirade because I had just come into possession of one of those amazing pens with the clickable ink top that allows you to pick one of five colors to write with. That morning I had switched from blue to black to green and then to pink on my writing exercise, and Mr. Hairy Arms had caught me. He scolded me and threatened to take away Shirley. (Yes, my pen's name was Shirley.)

I was feeling on edge and was growing ever more concerned for Shirley's safety given the current topic of Mr. Hairy Arms' lecture. I could just see my prized pen sitting at the bottom of Mr. Hairy Arms' drawer of confiscated treasures. That drawer was full of fake dog poop, Elsas, Lego Friends, and other pens just like Shirley.

A freckled arm shot up from the back of the room. "Mr. Brown," a girl said.

Yes, his name was technically Mr. Brown, but if you had ever seen the length and thickness of the hair on Mr. Brown's arms you'd understand. It was one of those cases where a name did *not* tell you very much about a person. Brown is a nice, neutral color perfect for boots or pants. Mr. Brown's arms were covered in dark, mossy wool that made it difficult to discern what actually existed between his shoulder and his wrist. For all I knew, he could have been hiding mutant reptilian arms or possibly even robot appendages under all that hair.

"Yes, *Mizzzz* Copper," Mr. Hairy Arms smoothly replied.

"Are there any circumstances in which the use of a pen is acceptable?"

Mr. Hairy Arms wiggled his mustache and scratched one of his woolly arms. "And *what* circumstances might those be?"

Mr. Hairy Arms also had a way of punctuating words like a CIA interrogator. He could say the word "how" in such a way that both asked a question and accused you of serious espionage crimes.

The girl with the freckled arms blinked. "Well, that is what I am asking you."

Mr. Hairy Arms' equally hairy eyebrows began to bounce in such a way that I knew my classmate had crossed the line. A copper color flashed across her freckled nose, and I could tell she knew it, too. I began to panic. I didn't want to see her sent to the principal's office just because she asked a simple question. Especially a question that I myself wanted answered.

"What she means," I said, rising to my feet and squeezing Shirley tightly in my hand, "what she means is ..." I squeezed Shirley once again and felt a drop of sweat form in the palm of my hand.

What am I doing? I asked myself. *I'm going to be the one sent to the principal's office.*

"What I mean," the girl said, "is that we used to use pens all the time in Germany. But now, for some reason, I should not use a pen. Why is that?"

Tori moved to Texas with her folks the summer before first grade. Her mom used to be in the military and they had lived all over. Her room is covered in postcards and snow globes and pictures from all of the places she has been. She's skied on snow-covered mountains and been on trains that drive all night and have bunk beds that you can sleep in. She's even eaten french fries that were covered in mayonnaise. That's right, mayonnaise on fries. She even liked it! *Gross!* But I digress.

"*Germany,*" Mr. Hairy Arms had replied. "Well, we are not in *Germany.* We are in *Texas,* and in *Texas—*"

"STOP!" Tori screamed at the top of her lungs.

Mr. Hairy Arms froze mid-sentence. Even his mustache stopped wiggling.

"I beg *your* par—"

"DON'T MOVE!" she yelled and slowly approached our teacher.

"*Mizzzz* Copper—"

"HE'S ON YOUR ARM," Tori hissed as she crept near Mr. Hairy Arms.

Mr. Hairy Arms' mustache began to wiggle so quickly that I thought it might fly right off of his face. "*Mizzzz* Copper, I don't appreciate ..."

Tori crept forward and pointed toward a particularly bushy patch near our teacher's elbow. "There. Right there," she said through her teeth, obviously trying to whisper.

Mr. Hairy Arms' eyes followed Tori's arm and finger. His eyes suddenly grew wide.

At once, Tori and Mr. Hairy Arms yelled. "LELAND!"

Leland, our class tarantula was nesting in the crook

of Mr. Hairy Arms' arm. The big, black spider looked perfectly happy and cozy all cuddled up in that giant nest of hair. There was no way to tell how long Leland had actually been on Mr. Hairy Arms' arm. All of that bushy, black hair was the perfect camouflage for a class tarantula that obviously had needed a change of scenery from his boring, old terrarium.

"AH, AH, AHHHHHHHHH!" Mr. Hairy Arms screamed in a high-pitched way.

I cupped my hands over my ears.

The kids on the front row joined in. "AHHHHHHHHH!"

Poor Leland looked confused and maybe even a little annoyed.

Tori moved forward. "Don't move. I'll rescue him."

"Rescue him?" Mr. Hairy Arms yelled. "Rescue *him*? What about *me*? Someone needs to rescue *me*!" Mr. Hairy Arms pleaded.

"Don't move," Tori commanded.

"AHHHHHHHHH!" Mr. Hairy Arms screamed and began to flap his arm like a bird.

"You'll hurt him," Tori said in a panicked voice. "You have to calm down!"

Tori loves bugs, even spiders. Tori spent time learning about the local bug populations in all the different places she lived by observing them in their natural habitat, taking pictures, and reading about them in books and on the internet. She told me once that moving a lot and leaving behind all of her friends had made her scared, but she had known that no matter where she went there'd always be bugs and scared spiders. So she found ways to make *them* her friends.

"AHHHHHH!" Mr. Hairy Arms continued to scream and wave his arms like a giant, furry bat. "AHHHHH!"

Kids in the back were starting to panic. Mr. Hairy Arms was whipping the entire class into a frenzy.

"It's gonna bite him," someone said.

"It can't bite through all that hair," said someone else.

"What if he flings it on your face?" another person interjected.

That statement seemed to be the hysterical tipping point. Mr. Hairy Arms began to flap his hairy arms quicker. Kids panicked and ran around the room, and Tori stood tall and brave trying to rescue poor Leland.

"Mr. Brown, you have to calm down if you want me to get him off you," Tori said.

Mr. Hairy Arms continued to scream and flap.

"He's airborne!" an observant kid yelled from the back of the room. "Leland's coming for your face!"

I swung my head toward Tori and Mr. Hairy Arms just in time to notice a scared, furry tarantula taking flight. Mr. Hairy Arms' flapping had flung Leland right out of his nest and toward the ceiling. His tiny spider body was flipping and flying through the air.

With the grace and agility of an all-star tee-ball player, Tori stood under Leland, caught him in two cupped hands, and gently placed him back in the terrarium.

"There you go, big guy," she cooed.

My mouth gaped open as I watched her so easily and gently handle the seemingly deadly spider. She smiled with her copper-colored lips and stroked his tiny head with her freckly fingers.

"You aren't scared of him?" I asked.

Tori shook her head.

Tori Copper was one of the bravest people I had ever met. Most of the other kids acted like Leland was a brown recluse or something. They were afraid Leland would bite them and send them to an early grave. They never took the time to understand anything about Leland and other tarantulas.

Leland was actually a gentle spider who liked to eat grasshoppers and mealworms and the occasional grape. He loved sleeping under his rock and watching Mr. Hairy Arms write on the white board. I had a theory that Leland thought Mr. Hairy Arms' arms were some sort of tarantula colony and watching those arms made him sentimental for his family.

Tori and I looked back toward the chaotic class. Kids were hiding under desks, Mr. Hairy Arms was running back and forth continuing to flap his arms, and Todd Brakefield was in the corner sucking his thumb.

"They're usually more scared of us than we are of them," Tori said.

I raised my eyebrows and pointed toward Todd. "Are you sure about that?"

Tori and I giggled.

"But weren't you afraid he was going to bite you?"

I asked.

Tori shrugged her shoulders. "No, he wasn't really showing any defensive signs like standing on his hind legs or showing his fangs. Beside, a tarantula's poison won't kill you. It feels like a bee sting mostly."

"You've been stung?" I asked.

Tori nodded her head. "Yep. And look, I'm still here."

I smiled at Tori. Not only was she brave, she was knowledgeable.

"I like bugs, too," I said as I watched Tori fill up Leland's food bowl and pet his little furry head. I knew then that Tori Copper and I would be friends for a very long time. "Do you have a best friend?" I asked.

"No. Do you?" Tori scrunched her freckled nose as she looked at my feet. "I think you forgot to put on matching shoes this morning."

I looked down. As usual, I was mismatched. I was wearing one fringed western boot and one strappy sandal.

I shrugged my shoulders and put my arm around Tori's. "If we're going to be best friends, there are a few things you should know about me. Starting with my footwear."

CHAPTER THREE

"Get him off!" Tori shrieked as a giant, green ninja kicked her freckly face. She twirled in a fast circle, wiping furiously at it.

I stared at her in shock. Okay, maybe it wasn't giant, or a ninja, but it was green and it *was* pretty darn big. But freaking out over a bug wasn't like Tori.

Every Tuesday night, Tori and I meet for bug collecting and pizza eating. It's a tradition we have had ever since bonding over Leland. Our Tuesday night bug nights allow us to explore our local bug population and eat our favorite pizza, Mr. Magali's Buy-The-Slice.

My dad and Tori's dad take turns driving us to Tuesday night bug nights. Tori's dad is a super-smart entomologist who works for the state's entomology lab. He is dispatched whenever there's a major bug emergency, like that time there was a tarantula stampede in West Texas or the time locusts swarmed a small town in East Texas. Tori's dad is in charge of understanding why major bug events happen. He's also in charge of reassuring people that the locust swarms are just a seasonal occurrence, not biblical.

Mr. Copper has tons of exciting stories about spiders and scorpions and snails. But tonight we had to settle for my dad. He's an entomologist, too, but he works at the university and just writes books about bugs. He doesn't have any stories about tarantula stampedes. All he talks about is how much he hates writing bibliographies and shows us his paper cuts. I don't think a real-life spider has even bit him.

"Is he an endangered species?" I asked. Surely the only reason why Tori would try to intentionally shoo away an awesome creature like the giant, green ninja was because she didn't want to accidentally squish him to smithereens.

Tori gave me a funny look. Her nose wrinkled up like a pug dog and her freckles seemed to pulse. "What are you talking about, Sera?"

I shrugged my shoulders. "I've never seen you pass on an interaction with a massive, awesome bug like that. I mean, it must be three inches long. Look at his wings; they're at least two inches by themselves. I bet he could fly all the way to Mexico in one night."

Tori rolled her eyes, and the grasshopper fluttered. It seemed to be insulted by her lack of interest.

"I'm just not into it today," she said, sighing loudly and laying her bug net on a bench.

"Okay," I said. "No pressure on the bug gathering. You must already be stocked up."

Tori sighed again and sat down on the bench with a loud flop. "I just have some things on my mind."

I looked at Tori. The corners of her mouth were

hanging on the side, like her lips had been melted, and her eyes were closed. Almost like she was in pain.

I sat down next to her. "Is Coach Cowell getting you down?"

Coach Cowell—our crazy, super-weird P.E. coach—was obsessed with dodgeball. I despise dodgeball! I can't stand getting nailed with those big, red balls. Give me kickball or a friendly game of soccer anytime.

Tori shook her head. "It's not Coach Cowell."

I frowned. "Is it Mrs. Dugan and her woefully inaccurate perspective on the chocolate versus vanilla debate?"

Mrs. Dugan, the school lunch lady, had been on a chocolate strike. You couldn't find a piece of chocolate or any chocolate-flavored product in a single square inch of our cafeteria. No chocolate milk, no chocolate ice cream, no chocolate pudding, not even a chocolate chip cookie. Quite frankly, I'd been afraid her behavior was pointing toward a serious break from reality. What kind of person doesn't like chocolate?

"No," Tori sighed. "You know I like vanilla better anyway."

I gasped. "How could you, Tori?"

Tori laughed. Her smile came back and happiness spread over her face like the rising sun. "Sorry, Sera."

Her face reminded me that it was the first time I'd seen her smile that day.

"What's got you down, Tori?" I asked, placing my hand on her shoulder.

Tori's smile faded away.

"It's my parents, Sera. They've been fighting."

"Again?" I asked.

Tori nodded.

I tried to understand why that would be making her upset. My parents fought all the time, and it never made me that upset. Not upset enough to give up on a perfectly good grasshopper.

My parents can't agree on anything. Dad's favorite author is Jules Verne while Mom prefers Ray Bradbury. Mom likes anchovies on her pizza, and Dad thinks it makes the pie smell fishy. Dad prefers straightforward quantitative data while Mom thinks there is much more merit to qualitative research. Just that morning I'd seen Dad erasing all of Mom's zombie movies from the DVR because he said it was giving him nightmares! But no matter how much my parents fight, I know that they're just two different people with different opinions. Their differences make them closer. Like the way Dad told me he was too scared to fly in an airplane until he met Mom. Now he flies lots of places. He told me that he would have never been able to see the world unless he had Mom.

"This is a different kind of fighting," Tori said. "This is bad fighting. With lots of screaming and crying."

I nodded my head, trying to understand.

"I thought everything would blow over after your birthday," I said. "I mean you said it usually blows over right?"

"Usually." Tori pressed her lips together and

closed her eyes. "But yesterday it got so bad that my mom left the house. She didn't come back home until this morning. She told my dad that they needed to have a long talk. I don't think it's a good talk that she wants to have."

"What do you mean?" I asked.

Tori paused. She passed her specimen jar back and forth between her hands. "Mom said they needed to talk about the future."

I watched her fiddle the jar around. "Well, that's good, right?"

Tori shook her head. "No, Sera."

My heart started to beat, and my chest felt all lumpy. Something goopy and gross inside was knotting up and making my stomach hurt. "What do you mean?" I asked.

Tori took a deep breath and pushed back her copper-colored hair. As usual, it was all wiry and sticking out in a thousand different directions. Tori was usually too focused on her bug collection to concern herself with frizz control. "My mom said she wants to divorce my dad."

"Divorce? Like not being married anymore?" I felt like I had been punched in the gut.

Tori nodded in a really rapid, jerky way. "Yeah. And my mom said she is going to leave."

"Leave?" I asked.

Again, Tori nodded in her jerky, weird way. "Yeah, like move out. Like she wouldn't live with us anymore."

I blinked. I wasn't sure how to process what I was

hearing. "I don't get it. You mean like move into a different room. Or like next door or something?"

Tori shook her head, in a slow, back and forth way. "No. Like she is going to leave and go away. Like, really far away. Far away and never come back."

It was my turn to shake my head. "No. That's not right. Because if she left, like, really far away and never came back, then you would never see her."

Tori blinked. "I know."

For the first time ever, I saw Tori Copper cry.

Tori wasn't one to cry. But these were really big tears. Tears with snot and chest heaving and shoulder shaking.

I felt helpless. I wasn't sure what to do.

"It's going to be okay," I said patting Tori's back, trying to avoid the snot bubbles that were forming in her nostrils.

"How do you know?" Tori said. "You don't know anything about divorce."

I was stunned. Tori was right. I didn't know anything about divorce. "Yeah, but," I stammered, "but I know lots of stuff about things."

"Things? What things?"

"Uh, science things," I replied confidently. "Yeah, I know about chemical compounds, and covalence, and how this whole universe works, so I'm pretty sure I can figure this out."

Tori sniffled and wiped her nose, popping those horrible snot bubbles. "I just want my parents to stay married."

I nodded deftly. "Then that's what we're going to

do, Tori Copper." I rose to my feet. "We are going to make sure your parents stay married."

CHAPTER FOUR

I'd had lots of ideas in my life, but none of them had involved the manipulation of romantic relationships. Granted, most of my ideas did involve manipulation—like the manipulation of photons and neutrons and acid/base solutions—but I had never directed people's feelings. As luck would have it, though, we were covering the periodic table that week in school and I was already way ahead on atomic weights and numbers so I was able to devote a lot of daydreaming time to various ways I could bring the Coppers back together.

The plan was simple yet brilliant: I had to create the most romantic situation that ever existed so that Tori's parents would fall back in love and call off the whole divorce thing. Thanks to my mother's expertise in all things romantic (and by expertise I mean her nonstop watching of super sappy, kissy-face, gross chick flicks) I had distilled the romantic mechanisms of dating into one extraordinary dinner that Tori and I would prepare.

I scanned my notebook, reviewing the major

constants, variables, operators, and functions that composed my romantic algorithm.

"Okay, Tori, let's go over this again," I said.

Tori nodded her head.

"Italian food?" I asked. For some reason, almost all romantic dates involved Italian food—and not just pizza.

"Check," Tori said. "I have several cans of pasta in the pantry, and we have a frozen lasagna in the freezer."

I nodded my head and checked the first major romantic component off the list. "Let's get the pasta on the stove and the lasagna in the oven." I cranked the oven to 500 degrees. Tori watched me as she used the can opener on the brightly colored cans of ABC pasta.

"Are you even reading the instructions, Sera?" she asked. "I think that temperature may be too high."

"Nonsense," I said confidently. "Those instructions are just mere recommendations. I need this to cook before your parents get home from work so I am just adjusting things slightly. It's math, really. I'm just accelerating the process."

Tori raised her eyebrows. "I'm not so sure about that."

I inserted the frozen lasagna—package and all—into the oven and gently closed the door. "Easy peasy."

"If you say so," Tori said, dumping a third can of ABC pasta into the pot on the stove.

I looked at my list again. "What about dessert?"

Tori pursed her lips. "I didn't think about dessert. But we do have ice cream. Will that work?"

"No. No way. Not romantic enough. We need something with chocolate and fruit."

"Chocolate and fruit?" Tori asked.

"Yes." I pointed out an entry in my notebook. "See? It says right here, 'chocolate covered strawberries'. I saw at least five instances during my romantic comedy research."

"I don't have any strawberries. I think we are out of fruit. I ate the last banana this morning."

"Okay, not a big deal. We can improvise." I opened the pantry door. "Do you have anything fruit flavored?"

Tori stood beside me and began to pull items from the shelves. "Lime jello. Fruity marshmallows."

"Perfect," I said. "Now we just need some chocolate."

"Hmmm ..." Tori dug around some more. "What about these?"

"Perfect!" I said as I grabbed a package of Oreos from her. "Everyone likes Oreos!"

"Oh, and I have wine, too!" Tori pulled out a bottle from the pantry that read "cooking sherry".

"Great!" I scanned my notes. "Next is music, but it has to be romantic music."

"Like what?" Tori asked.

I flipped my pages back and forth. "I must not have found that information."

"Then what should we play?"

I thought for a moment. "It seemed like going to

operas was a major thing to do, so probably opera music."

"I'm on it," Tori said. "I'll download some right now."

"Perfect!" I scanned the kitchen, proud of our handiwork. The ABC pasta was bubbling on the stove. The flowers we'd clipped from Tori's mom's garden sat in the middle of the table, on top of the bed sheets we'd used in place of the crisp linen tablecloth we couldn't find.

"So silky," I said as I ran my hand over the sheet.

"What's left in the formula?" Tori asked as she swiped and pointed on her screen.

"Let me look." I scanned my grand formula:

$[F]$the perfect romance = {(Italian food*wine) + (Δ)flowers + (X)}/πr^2mushy make-out face.

"We have the food, the flowers, but I'm not sure how we can artificially simulate the mushy make-out face."

"Mushy make-out face?" Tori inquired.

I nodded. "Yeah. The mushy-make out face. You know that really stupid vacant look couples seem to get right before they start smooching. It's all dead shark eyes and zombie mouth kind of a thing."

"Right," Tori said. "Mushy make-out face."

"That seems to be the biggest weakness in the entire formula. I just don't know how to make that happen."

"What about a gift?" Tori asked.

"Oh, you don't have to get me anything. Your friendship is enough."

"Not you!" Tori said. "Them!"

I scratched my temple. "I see where you're going with this. Yes. That just might work."

Tori nodded.

"But what do we get them?" I asked.

"Something they like. Something that will put them in a good mood."

Tori and I began to rummage around the kitchen. We opened drawers and cabinets. Tori opened the drawer next to the sink and was almost crushed by a cascading tower of Tupperware dishes.

"I was supposed to organize that last week," she said.

"Maybe that could be the gift," I suggested. "The gift of organization."

Tori shook her head. "No. It has to be something more than that. Something that gives them both those dead shark eyes."

I nodded.

"What do they like?" I asked.

Tori shrugged. "Outdoor stuff, I guess."

"Two-cycle engine oil!" I exclaimed.

"What?" Tori asked.

Two-cycle engine oil was something that my dad always asked for. Every time my mom asked my dad what he needed he told her that he needed two-cycle engine oil. *Or maybe that was just the last time we went to Home Depot*, I thought. Never mind, we needed a gift, and we needed it now!

"Let's go the garage," I said. "I bet we can find it there."

"Alright." Tori led the way into the garage. She opened the door and turned on the light.

"Why do garages always smell bad?" I asked, waving my hands in front of my face.

"What does it smell like?" Tori asked.

"Oil," I replied.

Tori began scanning the shelves. "What's it called again?"

"Two-cycle engine oil."

"What about engine oil?" Tori asked.

"No. No good," I replied. "It has to be two-cycle."

"Here it is!" Tori raised a small, red bottle in the air.

"Woohoo!"

"Grrrrrrrrrrr," the garage door moaned as it rose.

"Someone's here!" Tori yelled.

"Quick, we have to get to the kitchen!" I screamed.

Tori and I ran back into the kitchen. We rounded the corner and nearly slipped on the tile.

"Run!" I yelled.

As we entered the kitchen, a slight gray fog bloomed near the oven.

"Do you smell that?" Tori asked.

I struggled to catch my breath. "It's probably," I panted, "just the oil."

Tori sighed.

"Well hello, girls," Mrs. Copper said, laying her purse on the table. "What are you up to?"

"Do you smell smoke?" Mr. Copper asked as he entered the kitchen. "Tori? Serafina? What on Earth are you girls doing?"

Tori and I stood shoulder to shoulder and smiled broadly.

"We made you a romantic dinner," Tori announced.

"Romantic dinner?" Mr. Copper said. "What's the occasion?"

Tori looked at me quizzically.

"Uh ... well ... to celebrate love, of course," I said.

"Of course!" Tori said with a flourish of her hands. "Please sit down. Let me pour you some wine."

"Wow," Mrs. Copper said. "Wine?"

"That's right," Tori said pulling out two plastic tumblers. "I just couldn't find the wine glasses so I hope you don't mind."

Tori's mom shook her head. "Not at all. This is very nice, ladies."

Tori's dad sat down. "I think whatever you have in the oven may be burning."

I took a deep whiff of the air. *Eek!* I quickly pulled on two oven mitts and opened the oven door to find something that looked less like a lasagna and more like a piece of burning charcoal. "Tori. I think we might need a fire extinguisher!"

"What!" Tori shrieked.

"Oh my goodness!" exclaimed Tori's mom.

Mr. Copper jumped up. "Let me take a look at this," he said as he ducked his head toward the oven. "She's right. Get the small one from under the sink."

Tori's mom ran and retrieved the extinguisher. Then, with a few frantic hand motions, she was spraying white foam all over my romantic lasagna.

"Is it out?" Tori's dad asked, coughing.

Tori's mom hacked. "I think so."

"At least we still have pasta," I said, grinning sheepishly.

Tori's parents sighed and sat back at the table.

"I told you that was too hot," Tori hissed.

"Shhh," I reprimanded. "Just serve the pasta."

Tori's mom choked again, only this time it was on her drink. "Did you girls salt the wine?" she sputtered.

Tori and I looked at each other.

"I didn't. Did you?" Tori asked.

I shook my head.

Tori's dad took a drink and winced. "I think I know what the problem is." He stood up and picked up the bottle Tori had found. "Cooking sherry."

Tori's mom nodded her head knowingly. "I see."

"Is that not right?" I asked.

"Cooking sherry is for cooking only, not drinking," Tori's mom answered.

I furrowed my brow.

"It comes salted."

I pursed my lips.

"Wine and salt usually don't mix, Serafina."

I shrugged my shoulders.

"Would you guys like some romantic music? Maybe some opera?"

Tori's parents looked at each other and smiled.

"Why not?" Tori's dad responded.

Tori fiddled with her smart phone as I served the ABC pasta. "Bon appétit."

"Yum!" they both said.

Tori finally got the music on.

"Is this the real life? Is this just fantasy?" a male voice blared as we all covered our ears with our hands.

"Turn it down!" Tori's mom yelled.

"I'm trying!" Tori screamed as she pressed the screen.

"Wait! They're speaking English!" I screamed over the blaring words. "I thought opera was supposed to be in Italian."

"The description said it was rock opera!" Tori yelled.

"Too loud!" Tori's dad said over the music.

"I think I got it." Tori wiped her forehead as the volume of the music came down to a normal level.

"Wow," Tori's dad said as he rotated his finger in his ear. "I love Queen but not on concert level."

"Bohemian Rhapsody," Mrs. Copper mused. "One of my favorites."

I clasped my hands to the side of my head. "We totally messed this up. This was supposed to be a really romantic dinner for you both."

Mr. and Mrs. Copper smiled and looked at each other.

"Who said this wasn't romantic?" Mr. Copper asked.

I breathed a deep sigh of relief and gave Tori two thumbs up. She smiled in return and busied herself with preparing dessert while I returned to the oven to see what part of the lasagna could be salvaged. I figured if I could scrape all that fire-suppressing foam off the top, we should be good to go.

I wrestled off the black box to reveal the semi-frozen and semi-charred lasagna. *Perfect. They can just eat around all the charred parts.* I balanced the lasagna on my mitts.

"Girls," Mrs. Copper said from the table. "Are these my good silk sheets?"

"Uh," Tori said.

"Is that a yes?" Mrs. Copper queried.

"I guess ..." Tori began as I plopped the black-and-red casserole onto the table. Cheesy, tomatoey liquid sprayed onto the make-shift tablecloth.

"Bon appétit!" I declared.

"My sheets!" Mrs. Copper cried. "They're ruined!"

Tori and I looked at each other in a panic.

"Flowers. Look, Mom! We have flowers." Tori pushed the vase of roses and dandelions in front of her parents.

"Holy cow!" her dad yelled. "That's why I can't stop itching."

My eyes focused on Mr. Copper's right hand. It seemed to be super-sonic scratching a horrible looking red rash on his neck.

"I'm allergic to dandelions, girls. We have to get this out of the house immediately."

Tori looked stricken. "This is not working."

I began to nervously pace and flip through my notebook. "The gift," I said. "Give them the gift."

Tori quickly retrieved the bottle of oil and ran toward the table. As she ran, she tripped on a frozen piece of lasagna that must have fallen onto the floor. The bottle went flying from her hand, all slow-motion style, and hit the table.

"Nooooooo!" she screamed as the cap came loose and oil sprayed all over the silk sheet and her parents.

Her parents' faces froze. Their mouths formed into small, angry "O"s. It wasn't exactly the kind of face I had been going for. I had been hoping for more dead-shark-eyed, open-zombie-mouth, about-to-kiss kind of expressions. Instead, we had more of a "you ruined my silk sheets, committed arson, and will now be grounded for life" kind of looks. Not good.

"We have dessert," I said in a small voice.

"Shut up, Serafina," Tori grumbled.

CHAPTER FIVE

Friday night, my brother Apollo and I were sitting in the living room watching E.T. when the phone rang.

"Don't answer that, Sera. It might be Mandy," Apollo said.

Mandy is Apollo's super-lame girlfriend. I made a barfing noise and kept watching the movie while he left to go answer the phone.

I have a love-hate relationship with my brother Apollo. Seriously. He's in high school, which apparently makes him smarter and more qualified to be the kid in charge in our house. I've repeatedly demonstrated my superior reasoning and logical skills to my parents, yet somehow they always leave Apollo in charge!

Apollo is gross. He has giant smelly feet and plays basketball in the house. He also hates bugs, unknown chemical substances, and never wants to discuss theoretical physics. It's like we have nothing in common!

Apollo watches me and our baby brother, Horton,

a few nights a week. Mom and Dad had decided they wanted to go watch the new zombie thriller at the movie theater that night, and so I was—once again—stuck with Apollo in charge.

"Hello," Apollo said as he picked up the receiver. He said it in the tone of voice he uses when he's trying to be cool. I rolled my eyes. "Oh, hello, Mr. Copper. I'm afraid my parents are at the movies right now."

Mr. Copper? Why is he calling Mom and Dad? I stopped watching the movie and turned to watch Apollo.

"Yes, sir. I'll try and send them a message. I'll have them call you back as soon as possible," Apollo said and hung up the phone. He pulled his cell phone out of his pocket.

"What did Mr. Copper want?" I asked.

"Hold on, Serafina. I need to get a hold of Mom and Dad." Apollo tapped something out on his screen.

I rolled my eyes.

About a half an hour later, Dad and Mom returned home with Tori. She had a packed bag and her bug jar.

"Tori!" I shouted as she came through the door. "What are you doing here?" Thinking that might not have come out quite as excited as I meant it, I added, "Not that it isn't great to see you, but I just wasn't expecting—"

Tori smiled and hugged me. "I get it," she said. "And I get to stay with you this weekend."

"That's awesome," I said. "I really thought after that dinner you were never going to be allowed to have a social life again."

Tori shrugged.

Mom and Dad smiled.

"Alright, ladies. Why don't you get settled into Sera's room, and we'll make some popcorn," Mom said. "I'll bring it up to you guys in a few minutes."

"Sweet!" I yelled as Tori and I raced up to my room.

Tori didn't say much as she placed her bug jar on my dresser.

I sat down to fire up my computer. I had to check on the meteor and comet status of the sky. With Tori at my house, we could have an all-night, sky-watching party.

"Is the Leonid shower going to be active tonight?" I asked Tori. "I thought I heard something about it from Roger." Leonid is the super-awesome meteor shower that occurs when the Tempel-Tuttle comet comes into Earth's orbit.

"I'm not sure," Tori said as she wrenched off her rubber boot.

I watched her pop the other boot off as my computer roared to life. "Why do you always wear rubber boots?" I asked. "It's not even raining."

"Why *don't* you wear rubber boots?" Tori asked.

I shrugged my shoulders. "I'm not a fan of non-breathable material. My feet kinda sweat. Besides, I can only find one of my rubber boots."

"Well," she said, placing her hands on her hips, "I'm always prepared in case of a pop-up thunderstorm. What are you going to do if you get caught in the rain?"

Just then there was a knocking at the door.

"Come in!" I yelled, expecting to see Mom or Dad. Instead, standing there all dopey and annoying was Apollo. "What do you want?" I sighed.

Apollo ignored me and brought a giant bowl of popcorn into the room. "What's up guys?" he asked.

"We're discussing the merits of always being prepared for the weather," Tori said.

Apollo gave her a strange look and handed her the bowl. "Here, Mom wanted me to bring this to you."

"Great," I said. "Now if you don't mind, we're planning our activities for the night." I made a motion toward the door.

Apollo, too oblivious for his own good, ignored me. "How're you doing, Tori?" he asked.

I sighed and stood up. I began to push Apollo toward the door. "She's just great. Now please exit my lair."

Apollo sidestepped me and sat on my bed.

"Oh great," I said, throwing my hands in the air. "Now I can't get rid of him."

"Tori," he said, completely ignoring me, "I wanted to let you know that my best friend's parents also went through a divorce. So if there is anything you need to talk about, I'm here for you."

Divorce? Divorce? I thought, my mind racing. I looked at Tori's face and then back to Apollo.

Tori's mouth was a small slit. Her lips were white, like she was pressing so hard all the blood had been squished out. "Thanks, Apollo," she said, her voice cracking.

"What's going on?" I asked. "I thought you said they were just talking about divorce."

Tori nodded her head. "They were just talking about it until today. They had a giant fight last night, and this morning my mom was gone again. After dinner, they sat me down on the couch and told me that they were getting a divorce and wouldn't be living with each other anymore."

"I'm sorry, Tori," Apollo said. "That must be really hard."

Tori nodded her head and made her lips white again.

I looked at Tori and tried to think of something nice to say. *What exactly could I say?* I had no idea how she must be feeling. I knew that if I were in her shoes, I would be devastated.

I love living with both of my parents. They're both super cool and do lots of things with my brothers and me. Last year, when my mom was pregnant, we all went to Disney World and Dad rode all the rides with us while Mom scoped out the map. She wasn't allowed to ride the rides, given the fact she was almost as big as a small elephant. Mom wasn't even mad about it. Even when we all rode the teacups together and kept waving at her while she watched. She was an amazing sport about it and went and got us cotton candy.

Apollo said, "Tori, you must be confused. But I just want you to know that when parents divorce, it's not about the kid. It's about the grownups. It's not that they don't love you anymore, it's just that they can't make their marriage work."

"What does that even mean?" Tori asked, her voice crackling a little.

Apollo considered the question with a tilt of his head. "It's like in basketball," he said spreading his arms wide and rising to his feet. "Sometimes the best defense is zone. And sometimes the only thing that works is man-on-man coverage. Especially when you're playing those really fast kids from College Station High and you need to be in their face every second ..."

He was in full-on Apollo-Knows-Everything-About-Basketball mode. I lost track of what he was saying when he began to talk about alley-oops and the merit of understanding the fundamentals of a lay-up.

"Apollo!" I hollered. "You're not making any sense."

Tori giggled.

Apollo and I looked at each other and grinned. *At least we made her smile.*

"Listen, Tori. I don't know what Apollo is trying to say, but I know he's trying to make head and tails of this situation." Apollo shrugged in reluctant agreement. "He means well." He nodded his head. "What we need to do is to use something that we can both understand to help you out."

Tori nodded. "You mean science?" she asked.

"Exactly!" I shrieked.

"Kids!" Dad yelled from downstairs. "Try and keep it down. We're putting Horton to bed."

"Oops!" I said as Tori and Apollo laughed.

"Way to go, Serafina," Apollo said.

I punched him in the arm.

"So what I propose is that we use all of the scientific tools at our disposal to help you understand why your parents are getting divorced."

Tori nodded her head vigorously. "That sounds great!"

"We will use every basic tenet of the scientific method to explore and understand exactly what's happening and how to make it better." I pounded my fist into my hand. "Get your lab coat, Tori. We have work to do!" I raised my fist high in the air.

"Sera, I'm going to make you go to bed if you can't keep your voice down!" Dad called from downstairs.

"Quiet work," I whispered, pumping my fist in the air again.

Tori laughed.

CHAPTER SIX

I woke up Monday morning with a mission. I had to create a scientific task force/action committee to understand the basis for Tori Copper's parents' divorce. Furthermore, I had to figure out a way to make it stop.

Tori and I had totally messed up their romantic dinner, but I wasn't giving up hope. I figured we just needed a more systematic approach. One put together by the greatest scientific minds that existed at Sally K. Ride Middle School: Tori Copper, entomologist and rubber-boot enthusiast; Roger Penright, astrophysicist and part-time dog walker; Georgia Weebly, anthropologist and poet; and me, Serafina Sterling, leader and friend.

"Okay, guys," I said as my unofficial scientific task force sat down in the cafeteria to wait for the morning bell. "I've gathered you all here for a very important purpose. We have a major issue at hand, and I need all of the most important minds at Sally K. Ride working on the problem."

Roger chewed his pop tart slowly, moving his

mouth around and around like a camel. "Sera, I thought we were going to talk about what we did this weekend. I've got a hilarious story about the Smiths' three French poodles." He slapped his knee and snorted with laughter at whatever scene was playing through his mind.

Georgia squinted her eyes at him and sniffed the air. "Roger, what is that smell? Did you step in poo?" She pulled her fluffy, pink sweater over her nose.

Tori began to wave her hand in front of her face. "That must be French poodle poop. What did you feed them? Taco Bell?" She made a gagging sound and hid her face in her shirt.

"Guys," I said. "Forget about the poop. We need to focus." At that moment, the smell hit my nose with a force. I began to gag.

Roger checked both of his shoes. "Yep," he said. "There it is." He pointed to one of his high tops.

"Gross!" Georgia said with her tongue sticking out in disgust. "My eyes are watering."

"Sorry," said Roger. "Hazard of the job." He got up and went toward the bathroom, jumping on one foot all the way.

Tori watched him, laughing uncontrollably and gripping her stomach with both hands.

"Please focus," I said, trying not to barf. "We don't have much time."

Tori's giggles died down.

"What is it, Sera?" Georgia asked. "Please don't tell me we have to analyze the bottom of Roger's shoe."

We all began to laugh wildly.

"Shh!" Mrs. Dugan yelled from the corner of the cafeteria. "You girls need to keep it down!"

We covered our faces with our hands and laughed again.

Roger finally came out from the bathroom with a long trail of toilet paper attached to his shoe. "I think I got it all," he announced proudly, clearly oblivious to the toilet paper trailing his red Chuck Taylor.

I shook my head and tried to regain focus. "Sit down, Poodle Poop." I pointed toward his chair.

"Poodle Poop!" Tori laughed. "Classic!"

"Poop jokes are the best you have?" Roger asked. "What's next? Booger jokes? Maybe some good ol' knock-knock classics?"

Tori and Georgia giggled. I huffed at Roger's derision.

"Refocus, guys," I commanded. "We have a major agenda item that we need to address. Tori needs our help." I looked at Tori and waited for her signal.

Georgia and Roger looked at Tori.

"What's the matter?" Georgia asked.

"Are you okay?" Roger asked, concern growing in his voice.

Tori nodded her head. "I'm fine. And it's not technically my problem. Well, I guess it's kinda my problem, but really it's a problem between my parents."

Roger and Georgia furrowed their eyebrows in confusion.

Tori gulped. "You see, my parents told me this weekend that they're getting a divorce."

Roger gasped. Georgia covered her mouth with her hand.

"I'm sorry," Roger said.

"Me, too," Georgia added.

Roger looked at me. "I don't understand. How is this a scientific problem, Sera?"

"Well," I said as I looked at both Roger and Georgia who seemed to be asking the same question. "Science is a tool. Right?"

Both Georgia and Roger nodded their heads slowly.

"Science is a method that helps us understand things," I said.

"Yes, but," Roger said with a little apprehension in his voice, "where are you going with this?"

"Well," I continued, straightening out my back. "Science has and will continue to answer some of the greatest mysteries of mankind. Why not this?"

"Because," Georgia said, "this isn't a scientific mystery. This is a parental mystery. Parents do things we don't understand all the time. Like when my dad sold our minivan and bought a motorcycle. Then he grew his goatee really long and braided it and put pink barrettes in it."

Roger, Tori, and I all looked at each other.

"It's a long story," Georgia said waving her hands, as if to dismiss her previous statement. "The point is, I don't think science can help Tori make her parents stay together."

"Why not?" Tori asked rising to her feet. "I'm willing to try anything. Something has to work. Why not science? Science has cured diseases and discovered new species and explained how the universe began. Why can't it help my parents?"

Georgia and Roger looked at each other.

"I see your point," Georgia said.

"Me, too," Roger added.

"So can we count on you guys?" I asked.

"Of course," they said.

We all smiled.

"Thanks." Tori smiled her wide, copper-lipped smile. "I know that with you guys on my side, we can figure this out."

CHAPTER SEVEN

Scientists have a system that must be followed no matter what. This system—an adherence to the most basic principles of inquiry—binds the greatest minds from every known discipline, from astronomy to zoology and archeology to virology. We call it the scientific method.

Since I found out about Tori's parents, the scientific method seemed to be the only thing I could cling to. I mean, what exactly did I know about divorce? I knew that divorces happened. It was a reality of life. But I had no actual *knowledge* of divorce at all.

To me, divorce seemed to mean that everything doubled. Instead of one minivan, there were two. Instead of one Christmas, there were two. Instead of one house, there were two. *Right?* But that couldn't be all there was to it because having two of everything sounded kind of amazing and I knew enough about divorce to know it wasn't *that*.

What I needed to do was to stick to what I knew and that was science. I may not have been an expert on glitter or matching shoes—and certainly not on divorce—but I did know how to logically study a problem. For me to help Tori Copper I was going to have to turn to the scientific method. So that's exactly what I did. The steps were the same ones I'd walked through while studying dozens of other problems.

Step One: Define your purpose or question.

Step one is certainly the most important step. Understanding *what* you are asking is as important as understanding the phenomena you're observing. For example, if you're trying to understand the eating habits of the wild honey badger and pose the question, "Why do honey badgers prefer McDonald's over Burger King?" when in reality honey badgers actually prefer snails and turtles and other small creatures they can hunt (and they can't actually drive a car into town for a burger), you've missed the mark. Your question about honey badgers will never get an answer because they don't eat fast food! And not just because they don't have opposable thumbs.

You see, the question is what drives the entire scientific process of observation and data collection. You should always spend lots of time in research and contemplation to develop a most-righteous question or purpose. A good question gets a good answer.

Step Two: Conduct background research.

People always think that scientists spend lots of time just being cool in the lab, looking into microscopes, mixing noxious chemicals, and observing the wild honey badger in its natural habitat. In reality, scientists spend a lot of time understanding what other scientists have done in the past. My teacher, Ms. Wooly, says you "shouldn't reinvent the wheel". I think what she means is that there have been lots of great scientists who came before us who conducted some pretty solid work and paved the way to understand lots of things. Because of such pioneers, we already know a lot about the natural world. So in order to have solid and sound research, it's important to understand what other scientists have already discovered. In other words, you have to research and understand the background of your topic.

There are lots of places where you can do this. The library, for instance, has hundreds and sometimes thousands of books about every kind of topic you can imagine. Honey badgers, reptiles, the solar system, mucus (otherwise known as boogers) ... the list goes on and on.

The internet can be a great resource too, although no one is out there checking to make sure the information is accurate or reliable. Anyone can create a website. I mean Apollo actually made his own fan page! There are like a hundred pictures of him playing basketball on there. Barf a thousand times!

You can also ask experts like teachers and other scientists. I always talk to Mom and Dad before I start a project. Even though they can be lame and make me go to bed at 8 p.m. even though I have proven myself to be responsible and amazing and therefore should be able to stay up as long as I want, they actually do know a lot.

Step Three: State your hypothesis.

Hypothesis is basically a really fancy word for educated guess. That's right, science starts with guessing! Remember all those times your parents and teachers told you not to guess? Well, *guess* what? When you're exploring a scientific topic in depth, it's important that you start by making a *guess.* The hypothesis (or your well-thought-out and educated guess) is a statement in which you predict the outcome of your experiment. For example, if you really are committed to the subject of wild honey badgers and their culinary preferences, a testable hypothesis might be: *Wild honey badgers do not like to eat McDonald's or Burger King because they are wild animals and don't have a credit card or a mom who can drive them to get chicken nuggets when the craving hits.*

Okay, I know I'm being pretty silly with the whole honey badgers eating fast food thing, but the point is that these examples help illustrate how the scientific method works. Because after all, science helps us understand!

Step Four: Experiment Time!

Steps one through three usually involve a lot of reading and thinking. Sometimes that process can get super tedious and maybe a little boring, but it's what helps us become solid researchers. But step four is when the reward comes! Step four is when you get to throw on your lab coat, mix up some chemicals, and start making science happen! It's the experimental stage.

Experiments come in lots of shapes and sizes. Some scientists work in a lab where they have specialized tools for observing their variables, like microscopes and test tubes and specialized sensors or even the Large Hadron Collider. (In case you don't know about the Large Hadron Collider, aka the single most amazing piece of lab equipment ever, it's a giant particle accelerator that was built underground. In other words, it's a giant underground tube where physicists make particles go at insanely high rates of speed and then smash them together to make an intense collision! These collisions provide super-important information about particles that scientists previously had only been able to theorize or make guesses about.)

There are also scientists who do their experiments in nature, like biologists and zoologists or entomologists like my dad. Science is all about understanding the natural world. Sometimes, nature won't fit into a lab no matter what we try, and the science chicks and dudes of the world must go out into nature's territory to study it. Anywhere you find a question, you can create an experiment.

<u>Step Five: Analyze the data.</u>

Data, or information, is around us all day. We get information from the books we read. We get information from the labels of the food we eat. We even get information about how well we're doing in school on our report cards. (For the record, I usually get all A's, but my stinking Spanish class has been eating my lunch. I guess I can't *habla the español* all that *muy bueno* after all.) The data that's created from the scientific method—the *experimental data*—is the information we get from asking our question about the world and testing our guess about the answer.

Data comes in lots of shapes and sizes. Some data can be numbers, like how many bites it takes to eat one slice of Mr. Magali's giant pepperoni pizza. Mr. Magali's pizza has a four-foot diameter, that means the pizza is four feet across, and it usually takes me 32 bites to eat an entire slice. Tori can't eat a slice in fewer than 40 bites. But astrophysics enthusiast Roger Penright has eaten an entire slice in fewer than 15!

Numbered data is usually called *quantitative data* and describes scientific phenomena in numeric form. Like how much something weighs, how many inches tall it is, or how many times a person sneezes when they walk into Apollo's super dirty, super nasty room.

Data can also consist of words or feelings. This type of data is called *qualitative data*. Studying people and how they think or act sometimes means you can't use numbers or other traditional ways of measuring things. When a ruler and scale won't actually measure how you feel about something, you have to use words. An example of qualitative data can be your feelings about which flavor is the best ice cream flavor: chocolate or vanilla.

Step Six: Drawing a conclusion.

Once you have asked your question, done your

research, created your hypothesis, tested your hypothesis, and collected your data, you can finally start to answer your question. This final step is where you *draw your conclusions* or understand if your theory was correct or incorrect.

Sometimes your theory is correct, and sometimes it isn't. Either way, just by answering the question, you have learned a little more about science and made a meaningful contribution to the scientific world. That may seem strange to say but disproving a theory can tell you as much as proving a theory. Like when those Greek scientists from a really long time ago actually thought that the world was flat. Seriously! Real-life scientists thought that the world was flat and that if you sailed far enough you would just fall off the edge of the world. Eventually, some mathematicians like Pythagoras figured out the world was actually round. They disproved an existing theory and learned a little (okay, a lot!) more about the world.

CHAPTER EIGHT

"I've been thinking, Serafina," Tori said on Friday morning as she looked at me over our class scorpion's cage.

"Yeah, what about?"

"It," Tori said in a faint whisper.

"That killer clown movie?" I asked, watching Bartholomew crawl around. "I don't think you should watch that. My dad is still having nightmares."

Tori shook her head. "No, Sera. *Iiiiiiittttt.*"

I scrunched my nose in confusion. "What's *It?*" I asked.

"The deeeee ..." Tori took a deep breath. "The deeeeee ..."

"The divorce, Serafina. She's trying to say divorce," Georgia said.

I spun around to see Georgia standing behind me with her arms folded over her chest.

"I didn't even know you were back there," I said.

Georgia pushed her pink glasses up her nose. "I'll add that to the list."

"The list?" I asked.

"Nevermind," Georgia sighed. "Listen, ladies, we need a plan. The clock is ticking. If we are going to do something about this situation we must act fast."

Georgia waved Roger over to Bartholomew's cage.

Roger reluctantly stood and came over to the group. "I'm trying to finish my math homework. What do you want, Georgia?"

"You should have completed that last night, Roger Penright," Georgia hissed.

"No fighting," Tori scolded.

Georgia nodded her head. "You're right, Tori. Roger is just wasting my time."

Tori rolled her eyes. "I'm serious, Georgia."

Sobbing from the hallway interrupted us. Ms. Wooly jumped up to see where it was coming from. She walked her quick-teacher walk to the door, opened it, and poked out her head.

"What in the ..." Ms. Wooly said. "Charles?"

The sobbing grew louder—and sadder. I figured whoever was crying must have had something really horrible happen to them, like a deep puncture wound or a few missing teeth.

"Charles," Ms. Wooly said again. "Please come in here. Are you okay?"

We all stood on our tippy toes to see who "Charles" was.

"Mr. Brown!" Georgia gasped.

Mr. Brown, AKA Mr. Hairy Arms, was holding his wet face in his hairy hands and sobbing.

"Mr. Hairy Arms!" Tori and I said at the same time.

"Oh, Charles! Come in and sit down. What's the

matter?" Ms. Wooly ushered him into the room and led him to her desk while patting his back.

"It's Leland!" Mr. Hairy Arms exclaimed through his sobs. "LEEEELAAND!"

"Leland?" I asked.

Mr. Hairy Arms pulled his face from his hands. "Yes, Leland." He continued to sob.

"What's wrong with Leland?" Ms. Wooly cooed.

Mr. Hairy Arms once again pulled his face from his hands. Two giant snot rivers ran down his face.

"Gross!" Roger said.

Mr. Hairy Arms wiped his nose with his super hairy fingers. "This morning, at 8:35 a.m., Central Standard Time, Leland crept off this mortal coil."

My friends and I gave each other confused looks.

"'Mortal coil'?" Tori repeated.

"I am so sorry," Ms. Wooly soothed. She obviously understood Mr. Hairy Arm's cryptic language.

"*Why*!" Mr. Hairy Arms exclaimed. "Why do the good die young?"

"Ohhhhhh," Georgia said. "Leland must have died."

"What?" Roger asked.

Georgia rolled her eyes. "Come on, Roger. The 'mortal coil'? Only 'the good die young'? All the crying? Leland must have died this morning."

"Ohhhhh," the rest of us said at once.

"I just don't think I can go on without him!" Mr. Hairy Arms said through his shoulder-shaking sobs.

Georgia strode up to Mr. Hairy Arms with one of her pressed and monogrammed handkerchiefs.

"Here," she said extending the stiff hanky. "You need this."

Mr. Hairy Arms took the hanky and loudly blew his nose. "HOOOOONNNNKKK."

Georgia recoiled in horror. "Please keep it."

Mr. Hairy Arms sniffed. "Thank you."

Georgia nodded. "You know, Mr. Brown, the best way to honor Leland's legacy is to move on."

"Move on?" Mr. Hairy Arms asked in a shocked voice. "How could I possibly move on? I was with that spider for almost 10 years. A person can't just move on after 10 years!"

Georgia stayed defiant. "Of course you can. You simply must move on. You have to ensure the next generation of tarantulas, wolf spiders, jumping spiders, and other classroom friendly arachnids have a home in our public education system."

Mr. Hairy Arms sniffed and wiggled his mustache. "Now why is *that*, Georgia Weebly?"

Georgia stiffened and thought for a moment. "Because ..." She pushed her pink glasses up her nose.

Georgia has a lot of positive traits, but sensitivity is not one of them. She loves to give out advice, but she is rarely helpful. In fact, Georgia was banned from accompanying kids to the nurse's office because she kept telling sick kids to "walk it off". That's not so easy to do with diarrhea.

"*Because*?" Mr. Hairy Arms said in his CIA voice.

"Because ..." Georgia continued.

"Because"—I sprung to my feet—"you don't want the next generation of spiders to go the way of the boot, do you?" I was shouting at the top of my lungs.

"The way of the boot? What on *Earth* are you talking about, Serafina Sterling?" Mr. Hairy Arms asked.

I raised my eyebrows and arched my back just like any great professor would. "The way of the boot. You know"—I pointed to the cowboy boot on my right foot that accompanied the sandal on my left—"stomp, stomp?" I stomped my right foot to illustrate.

Mr. Hairy Arms gasped and grabbed his hanky. "*Mizzzz* Sterling. You're not saying what I think you're saying, are you?"

I nodded my head dramatically and stomped again. "Oh yes, Mr. Brown. The way of the boot. Just think about how many of Leland's poor babies and grandbabies and cousins and grandmas and arachnid tax professionals have gone the way of the boot." I stomped again, just to drive my point home.

Mr. Hairy Arms gasped again. "What exactly are you suggesting I do, *Mizzzz* Sterling?"

I held up both of my pointer fingers in another professorial gesture. Just to make sure Mr. Hairy Arms knew I was serious. "Spider education."

"Spider education?" Georgia repeated.

I gave Georgia the universal look for "you're not helping here" by scrunching my nose and puckering my mouth.

"Yes, *Miz* Sterling. I agree with Georgia. What exactly is spider education?" he asked.

I smiled my broad, fake smile that stretched out my cheeks but showed no teeth. It's the same way Mom smiles when I accidentally make chlorine gas and I ask her if she's mad. "I am so glad you asked. Spider education is a way to carry on Leland's legacy."

"How exactly am I supposed to *do* spider education, Serafina Sterling?" Mr. Hairy Arms asked in his most annoyed teacher voice.

This time I smiled a real smile. Wide cheeks and teeth showing so much I was afraid they might dry out. I was smiling because I knew that with Mr. Hairy Arms' help I could get Tori Copper's parents back together again.

"You are going to need spiders," I said coolly, not breaking my smile.

"Spiders?" Mr. Hairy Arms and Ms. Wooly asked together.

"Lots and lots of spiders," I said through my giant smile.

I looked at Georgia who was squinting her eyes at me. I wasn't sure if she was mad or possibly reading my mind.

"Why so many spiders?" Georgia asked in a reluctant drawl.

I shrugged my shoulders in an innocent way. "How else are you going to expose the next generation of kids to arachnids and their needs? They have to know spiders. They have to know big spiders and small spiders and poisonous and non-poisonous and blue and purple and glittery—"

Mr. Hairy Arms held up his hand in a "*please, stop!*" gesture. "I see where you are going with this, *Mizzzz* Sterling. But I have neither the time nor the ability to take care of so many spiders."

I smiled again. "Well, then you're in luck because Tori, Roger, Georgia, and I will all volunteer to take care of your spiders. Morning, noon, or night. We will be on call. You can just call us the spider sisters."

Roger shot me a look. "I mean siblings. Roger isn't a girl or anything."

"Sera," Roger moaned.

"I'm not sure I have time—" Tori began.

I quickly cut her off. "Sure you do, Tori. *Sure* you do." I shot her a knowing look and twitched my eyebrows.

Tori's eyebrows furrowed. "I do? Serafina, do you have an itchy face?"

"My face is fine," I said.

Mr. Hairy Arms stood and patted his eyes dry. "Then it's settled. I am getting lots and lots of spiders." He took a deep breath and grinned. "We'll call it Leland's Legacy and you wonderful children will be the caretakers."

"Serafina," Georgia whispered out of the corner of her mouth. "What did you get us into?"

I shook my head slightly.

Ms. Wooly must have seen us. "Kids, are you sure this is something you want to do?"

"Of course it is," I said. "This is a great *plan*. We were just discussing how we needed a *plan,* and I think this is the best *plan* we can come up with."

Ms. Wooly tilted her head. "Well ... if you all and Mr. Brown are okay with it, then I am, too."

"Then it's settled. Now we have that *plan* Georgia

was talking about," I said.

Georgia nodded her head knowingly. That's what I like about Georgia, not too much gets past her. "Okay, Serafina. We will go with your plan. But—"

"But?" Mr. Hairy Arms asked.

She smiled politely at Mr. Hairy Arms and then looked at me. "But it better work. Because that sounds like a lot of spiders."

CHAPTER NINE

Roger, Tori, and Georgia stood in front of me at seven on Monday morning. Everyone's arms were folded across their chests and they had very sour expressions on their face.

I smiled a sheepish smile. "Breakfast tacos?" I asked, holding up my white paper sack. "I have salsa."

Georgia swatted the bag from my hand. "I don't want your delicious breakfast, Serafina. I want answers."

I frowned and looked at the poor sack of tacos that had landed at my feet. The grease was starting to make the paper opaque. I frowned again.

Georgia looked at me sympathetically. "Oh, Serafina. I'm sorry." She picked up the sack and dusted off the greasy bottom. "You know I love breakfast tacos. I'm just so mad at you."

"Yeah," Tori said. "What's the big idea volunteering us for spider duty with Mr. Brown?"

"I could be sleeping right now," Roger added, yawning and rubbing his eyes.

"Look, guys," I said. "Don't you want to get Tori's

parents back together?"

My friends reluctantly nodded.

"Of course we do, Serafina. But how exactly is doing *this* going to get them back together?" Georgia asked.

"Just trust me," I said.

"Yoohoo!" Mr. Hairy Arms yelled, beeping the horn on his car. He was in his red convertible with the top down. "Today is a great day for spiders!"

My friends muttered, "Yay."

Mr. Hairy Arms pulled up to the curb and hopped out of his car. He opened the trunk with his key remote. "You are never going to believe the collection I have acquired. I have Mexican reds, wolfs, jumpers, grass"—he began to pull plastic cases from his trunk and hand them to our group—"fishing spiders, jumping spiders, crab spiders, barking spiders, garden spiders, and of course,"—he handed me three large, stacked tubs—"tarantulas."

"That's a lot of spiders," Roger mused, trying to balance all the boxes he was holding.

"Seventy-two spiders in all," Mr. Hairy Arms announced proudly.

"Seventy-two!" Tori exclaimed.

"Seventy-two," Mr. Hairy Arms confirmed with a proud smile and twirl of his hand. He snapped his trunk closed and gleefully walked to the front door of the school.

When we entered his classroom, we saw the brand new tanks and terrariums that the spiders were to live in. They glistened and glittered in the early morning light. Mr. Hairy Arms began flipping the lights on inside the tanks.

"That's a lot of food bowls," Roger murmured.

"And water bottles," Tori added.

"And fun!" Mr. Hairy Arms added, clapping his hand together. "I left a few things in the car. You kiddos make yourselves at home. We have a lot to get done today." He wiggled his fingers in the happiest early-morning wave a person could possibly muster on a foggy Monday. "Too-da-loo!"

The fluorescent lights of the tanks hummed as Tori, Roger, and Georgia stared at me silently.

I smiled. "Tacos?" I asked in a small voice. "I have salsa."

Roger shoved his hands into his pocket. "Serafina, this idea of yours better be good."

"It is. It is," I stated. "It's probably one of the best I've ever had."

"Let's hear it," Georgia said, pushing her pink glasses up on her nose and folding her arms across her chest.

"Ahem." I cleared my throat. "Well, if we are going to get Tori's parents back together, then we are going to have to go to extremes."

"Extremes?" Tori asked.

"I'm not sure I like the sound of that," Georgia said with a cautious tone.

"I agree," Roger added. "Serafina Sterling's extremes are usually ..." He thought for a moment. "... usually pretty extreme."

Tori and Georgia nodded their heads in agreement.

I pursed my lips. "And what exactly does *that* mean?"

"Like when you turned Apollo blue with all that silver dust," Tori said.

I shrugged my shoulders. "A simple miscalculation."

"Or when you tried to re-animate that frog we were dissecting in science lab by re-wiring Ms. Wooly's outlets, and you blacked out the school for three days," Roger added.

"I didn't hear anyone complaining when we got a three-day vacation," I retorted.

"Or that time you tried to create an automated diaper crusher for your mom and you got po—" Tori began.

I waved my hands. "Let's not talk about that right now. Quite frankly I am not in the mood to rehash some very simple inaccuracies. I mean, isn't that what science is all about? Trial and error?"

My friends made some muttering noises that either meant they agreed or they weren't quite awake yet.

"What I am proposing is a fool-proof plan," I said proudly. "No errors, no blackouts, and certainly no flying poo."

Tori snickered.

At least I'm making her laugh, I thought.

"Let's hear it then, Sera," Roger demanded.

"Alright then, Poodle Poop." I motioned for my friends to huddle up closely. They slowly formed a circle around me. I crouched down like I was a coach giving a pep talk in double overtime. "Alright," I said in my best conspiratorial whisper. "Here's the plan."

CHAPTER TEN

"So today's the day, right?" Roger asked, looking at his glowing smartwatch.

I nodded curtly to Roger. "That it is, Poodle Poop."

Roger smirked at me. "Well then, we better get to it. And Sera ..."

"Yes?"

"The poop jokes are so first grade."

It was my turn to smirk.

For almost five full days, Roger, Tori, Georgia, and I had been changing water bowls, filling water bottles, and putting out pellets, crickets, and mealworms (both alive and dead) for the 72 spiders that crawled and crept in Mr. Hairy Arms' room. We got to the classroom almost a full hour before the first bell rang and stayed almost a full hour after the last bell. We even went in at lunch! It was sun-up to sun-down work. I felt like I was working in a coalmine or something.

The spiders themselves had become legendary. Nearly every kid who attended Sally K. Ride Elementary *and* Middle School had come to see the arachnids. Eyes gaped as they watched the jumping spiders leap in their terrariums like eight-legged ballerinas. Mouths dropped as they heard the faint woof of the barking spiders. And Mr. Hairy Arms smiled as he watched kids of all shapes and sizes gush and coo at his vast spider collection. "Leland would be proud," he had said over and over through his wiggly mustache. "So, so proud."

"Okay, guys," Tori said, taking a deep breath as she replaced the last filled water bottle in the last tank. "Do you think this is going to work?"

Georgia gave me a nervous side-glance.

"Of course it will," I reassured Tori. "What could possibly go wrong?"

Tori shrugged. "I just keep thinking how freaked out Mr. Brown was that time Leland escaped. I'm just not sure how he will act when all 72 escape."

"How did he react?" Roger asked.

"He—" Tori began.

I interrupted Tori. "He wasn't *too* bad. Just a little jittery is all."

I thought back to Mr. Hairy Arms furiously flapping his arms in the air. I thought about the way all that furry moss on his forearms made him look like a bat when he was in motion. I also remembered that strange shade of scarlet he had turned. At the time, I kind of thought his face was going to burn off.

"Jittery?" Georgia asked. "My mom said he considered taking early retirement."

I shook my head furiously. "No, no, no. All lies."

As a point of clarification, Mr. Hairy Arms did

not retire after the Leland incident. That is simply absurd. He was obviously still employed as a first-grade teacher at Sally K. Ride Elementary School. He had, however, taken a *leave of absence* to deal with certain stress issues, such as having a giant spider on his arm for an unknown period of time and then flinging it across a first-grade classroom in a state of unbridled panic.

"And that he was medicated," Georgia added.

I had also heard the "medicated" rumor. However, I was told he was on heart medication for a slight arrhythmia that had developed after the incident. I was certain Georgia was referring to psychotropic medication, which was ...

"Nonsense," I replied. "You saw how much he loved Leland. He was crying like a baby for crying out loud! What makes you think that a few escaped spiders will send him over the edge?"

Roger, Tori, and Georgia exchanged skeptical glances.

"If you guys have a better plan, then let's hear it," I said with an irritated tone. "All I'm trying to do is get Tori's parents back together. Short of a miracle, this is the best I have."

My friends once again looked at each other. This time it seemed like they were all silently agreeing.

"So then, it's settled. We stick to the plan."

At that moment, it was the only plan I had. I had been racking my brain for several weeks about ways to stop the divorce but had come up short. The next best option involved intentional radiation poisoning, but I thought my mom might get really mad about having to claim that on our health insurance again.

Roger took a deep breath and looked at his watch. "The lunch bell should be ringing any second now."

"Then we better get to it," I said and opened the first spider cage. "Run free, little guy. Run free!"

A large, hairy tarantula named Doug looked at me with a confused expression.

CHAPTER ELEVEN

"Ms. Dugan!" I yelled. "Ms. DUUUUUGAAAANN!"

Ms. Dugan glanced up from behind the lunch counter. She pulled the gross looking towel from her shoulder and began to wipe her hands. "Serafina, if this is about chocolate again, I don't want to hear it."

I rested my hands on my knees as I tried to catch my breath. *I really need to work on my cardio*, I thought.

The lunch lady shifted in her white platform, lunch-lady shoes and folded her arms across her chest in irritation. "Then what it is, young lady?"

"Spppppp ... spppppppiiiii ...." I said in between my choppy breaths.

"Spy?" she repeated.

"Juuuuummm ... jummmmmmmppp ..."

Ms. Dugan cocked her head to one side. "What are you trying to say, Serafina? I can't understand you."

My cardio really is bad, I thought. *I can't even catch my breath.*

"Essssscccaappppeeee ... Theyyyyy gottttttt outttttt ..."

"SPPPPPPIIIIDEEEERRRRRS!" Todd Brakefield yelled from the back of the cafeteria. "THEY'RE EVERYWHERE! MILLIONS OF THEM!"

At once a cacophony of screams broke out in the lunchroom. Trays rattled as kids tried to find higher ground. Milk shot in the air like geysers as milk cartons were stomped underfoot. Square pieces of pizza flew about, leaving tomatoey streaks on cheeks like delicious slaps. Cookies broke in half. SNAP! Silverware fell to the floor in jangly rattles.

"SPIDERS!" kids shrieked.

"They're coming!" they screamed.

"Save me!"

I turned in Todd's direction to see a small clump of brown and black spiders slowing marching into the cafeteria. Leading the way was Doug the tarantula; a sour expression seemed to emanate from his furry face. Leaping behind him were the jumping spiders, followed closely by the Mexican reds, garden, barking, wolf, fishing, and crab spiders. It even appeared a few common household spiders—that had, no doubt, inhabited the school unknown and unseen—had joined the gang.

The spiders didn't seem ferocious. If anything, they seemed lost and a little scared. I immediately felt a lump in my throat. I was starting to doubt my plan. *What if the spiders got lost? Or worse yet, hurt?*

"Call Tori Copper's dad!" I yelled to Ms. Dugan. "He'll know what to do."

Out of the corner of my eye, I could see a pair of

flapping hairy arms.

"They're out!" Mr. Hairy Arms yelled. "What on *Earth* are they doing out?"

"Call Mr. Copper!" I screamed at Mr. Hairy Arms. "He'll know what to do!"

"AHHHHHHHH!!!!!" Mr. Hairy Arms screamed.

Hot fudge, I thought. *This is not going as planned.*

"Students and faculty, please remain calm," a voice came from the loudspeaker. "Remain where you are until you have been instructed otherwise."

I looked around the cafeteria just in time to see Georgia and Tori running toward me. A look of panic was spread across their faces.

"What's going on?" I asked.

"Pandemonium. Pure pandemonium," Georgia said in a breathless gasp.

I looked at both of them. "It wasn't supposed to get out of hand like this."

Tori shook her head. "I know. But apparently, Todd spotted the spiders and has been alerting everyone in the school."

"Alerting?" I asked.

Georgia nodded her head. "Yes, by screaming at the top of his lungs."

"Oh my goodness," I said.

"Yes, oh my goodness," Georgia added. "Now what do we do?"

"We have to get Tori's dad here," I said.

Georgia chided. "Sera, I think it's time to forget about the plan. The entire school is going insane because we let out all of Mr. Brown's spiders. I think we have bigger things to worry about."

"What can be bigger than Tori's parents getting a divorce?" I asked, feeling my own panic rise.

Tori patted my back. "Sera, your heart is in the right place. But your head is up your ..."

"Behind!!!!!" Roger yelled as he ran toward us.

"What?" I asked.

"They're coming up behind you!" He pointed toward the swarm of spiders.

Making their way through the cafeteria full of frightened kids was Doug and his small army of arachnids.

"Are there more of them?" Tori asked.

Before I could answer her, the loudspeaker squealed to life once again.

"We have declared a state of emergency at Sally K. Ride Elementary School. Your parents and guardians have been notified, as have the local authorities. Please shelter in place. The exterminators are on the way."

"Exterminators?" We all gasped.

"Exterminators!" Mr. Hairy Arms exclaimed between arm flaps. "Not my spiders!"

"What are we going to do, Serafina?" Tori pleaded.

"Yeah, Sera. Was this in the plan?" Georgia asked.

"What now?" Roger added.

I looked around at the mess that surrounded us. Kids who once had looked upon Mr. Hairy Arms' spiders with awe through the glass of tanks and terrariums were crying and groveling at the sight of those same spiders roaming the school free. Even though the gang of arachnids hadn't stung anyone or even spun a web, sheer terror was gripping my comrades. Snot and drool flung out of their red and pink faces. Teachers' dresses were covered in food that had been flung from lunch trays. Rivers of milk crisscrossed the floor. A faint whimpering seemed to echo across the school. I looked up and realized our principal must have accidentally left the loudspeaker on. It was *his* cries we were hearing.

"People get really worked up over just a couple of spiders," I mused.

"That's an understatement," Tori said.

I grimaced at her. "I don't know how it went this bad, but I'm sorry, Tori. I figured Mr. Hairy Arms would call your dad to help with all the runaway spiders and your mom would come to calm you down. And once their eyes met over the tarantula's cage they would fall back in love again. I never thought this whole thing would turn into such a terrible disaster."

Tori hugged me. "Your heart is in the right place."

I gripped my best friend tightly.

"Now," she said. "We have to rescue these spiders."

CHAPTER TWELVE

Mr. Hairy Arms; our principal, Mr. Bellows; and all of our parents stared at us in the totally judged-out way that only grown-ups can—eyes not blinking, mouths puckered, the words "you're grounded" being telegraphed into our brains.

We were sitting in Mr. Hairy Arms' classroom after 71 of the spiders had been successfully recovered. Doug the tarantula was still out on the loose and making a few of the adults jumpy. As an example, a dust bunny rolled over the toe of Georgia's mom's shoe and she stood up, began screaming, and stomped the dust bunny into the floor. Needless to say, everyone was on edge. It wasn't just Doug's unknown location; it was the fact that our plan had been exposed.

Under only minimal duress, Georgia had spilled her guts. She kept saying things about her "permanent record" and "Ivy League applications". Bottom line: Georgia Weebly would not be a good candidate for the military secret ops.

The adults didn't seem angry that we tried to get Tori's parents back together, but they were furious about our method. They claimed that we put teachers and students at risk by releasing Mr. Hairy Arms' arachnid army. I tried to explain that most of the spiders were non-poisonous and actually pretty gentle. That didn't seem to calm any of the red-faced parentals. They claimed that we squandered valuable emergency services and made our principal cry.

Mr. Hairy Arms shifted in his seat, crossed his legs, and leaned forward. "So let me get this straight."

I rolled my eyes on the inside. Rolling my eyes on the inside was a new thing I had invented. It was much more effective than rolling them on the outside and getting caught. Rolling your eyes on the inside involved a little practice and a lot of muscle control. I was resorting to the internal roll because this was only the 500th time Mr. Hairy Arms had questioned and re-questioned our plan.

"You *young* people thought that if you released 72 potentially dangerous spiders into our school that this would somehow force a reunion between Tori's mother and father?" Mr. Hairy Arms asked through his wiggly mustache.

"Yes, sir," I said with yet another internal eye roll. "As I mentioned a few *hours* ago, I thought that if we executed a controlled release of a few spiders that Mr. Copper would have to come to the school and help rescue the refugees. Then he would call Mrs. Copper cause Tori was gonna fake cry and act all scared for her life. Then when the Coppers saw each other they would look into each other's eyes and fall in love all over again."

I looked at Mr. and Mrs. Copper who were sitting next to Tori. They both blushed.

"Mr. Copper gets called out to emergency bug situations all the time. Why not this one? I thought that if we created an emergency situation, he would be forced to see Tori and her mom."

"That is so kind of you, Serafina. But the issue between Tori's mom and me isn't that simple," Mr. Copper said. "Besides, I don't really deal with escaped class pets, more like major bug emergencies that threaten people's lives or homes."

"I think a horde of poisonous spiders most certainly threatens my life. And makes my skin crawl," Georgia's mom said and shuttered for maximum drama effect.

I once again internally rolled my eyes. *It's like they didn't even listen when I told them most of these spiders are totally harmless.*

"*Mizzzz* Weebly, these spiders are harmless," Mr. Hairy Arms retorted.

"A person can still die of fright," she shot back.

Mr. Hairy Arms rolled his eyes, but on the outside. "My concern is and always has been for the safety of the spiders."

"Ahem," our principal, Mr. Bellows, cleared his throat. "And the safety of the students?"

Mr. Hairy Arms flapped his hands. "Well, of course, the students, too. But I was really afraid all my dear spiders might go the way of the boot."

"The way of the boot?" Mr. Bellows asked.

"You know? That stompy dance Miz Weebly

performed on that fur ball earlier."

Everyone nodded.

"Well it's clear the kids did something wrong and should be held responsible," my mom said. "But I do want to acknowledge that their hearts were in the right place."

Finally. A voice of reason, I thought. *Go, Mom!*

"I think the kids are having a hard time processing this whole situation," she added. "I'm not sure they understand what's going on here."

Now Mom was turning on me. It was the last straw. I felt my eyes roll, but this time on the outside. I sprang to my feet. "I *do* know what's going on here."

Everyone looked at me with shocked expressions. Mom's eyes widened. I saw her mouth, "No, Serafina!" but I ignored her. I felt it was my duty to let my voice be heard.

"What's going on here is my best friend's parents are divorcing and she doesn't like it. In fact, she hates it. She loves both of her parents and she can't understand why they just can't love each other like she loves them."

I looked at Tori whose face was flashing a copper color.

"I'm sorry, Tori. But someone has to say it." I rolled up my sleeves and began to point my finger wildly, just like I had seen Mr. Hairy Arms do on so many occasions. "I am sick of seeing my best friend cry. I care about her and want her to be happy and the only way for her to be happy is if you guys can be happy." I pointed my finger at Mr. and Mrs. Copper.

"So be happy!" I screeched and then stomped out of the room.

CHAPTER THIRTEEN

Tori, Roger, and I were hanging out by the swings when Georgia found us at recess. We had all been feeling pretty low because of the major scolding we received from our parents and the month of Saturday School we had to serve because of my failed spider plan. Doug was still on the loose and nearly the entire student population had developed arachnophobia. No one could turn a corner and see a stray gum wrapper or hairball without screaming and yelling, "KILLER SPIDER!"

"Guys," Georgia said, pushing her glasses onto her face and looking at her Lisa Frank notepad. "I've been brainstorming. You know, about the problem." She looked at Tori uncomfortably.

"Don't you think we should give it a rest? I mean, look what a mess this has gotten us into already," Tori said. "My parents are divorcing. They even have lawyers."

"Let's hear it," I said. "I don't give up that easily."

Tori shook her head.

"I mean it, Tori. You are my best friend and I am

going to make this work for you." I patted her back and gave her a hug.

"Well," Georgia said, taking a deep breath. "I think, since this is an issue with Tori's parents, who are people, we should be looking at this issue through the lens of anthropology. Anthropology, after all, is the study of culture." She tossed her hair in a self-important way, her bright yellow curls falling wildly down her back.

Roger rolled his eyes. "Here we go with Georgia thinking anthropology is the only answer to anything. I'm so sick of this academic superiority she has."

Georgia raised her eyebrows indignantly. "Well, you smell like dog doo-doo, Roger."

"Guys, guys," I said, waving my hand to end the fighting. "We have to be united if we plan to get anywhere. We can't spend our time fighting."

Roger lifted his shoe and pointed it at Georgia's face.

Georgia pointed her finger at Roger. "You better keep that poodle poop away from me!"

"Guys!" Tori shouted. "Please stop!" Her freckled face was bright red, like a ripe cherry.

We all looked at Tori with surprise.

"I can't take the fighting anymore. It's all my parents do all day!" Tori cried. Tori grabbed her things and spun on her rubber boot. In a few seconds, she was gone, stomping back into the school.

We watched her walk away, and then looked at each other. We were ashamed.

"Sorry," Roger finally said.

"Me, too," Georgia said. "I just wanted to help."

"Well, this isn't helping," I said, with my hands on my hips. "You guys are acting like a bunch of first graders."

They both nodded their heads.

After a few seconds, Georgia spoke. "Listen," she started. "We have to be on the same page if we plan to get anywhere."

"I agree," Roger said. "But I don't think we need to limit ourselves to just one field of study. There are lots of different approaches we can take."

We were silent for a few minutes thinking about the problem. On one hand, Georgia was correct. Not necessarily about anthropology being the only approach, but about trying to narrow our scope of study. On the other hand, Roger was correct, there were a thousand different ways we could approach this topic and still come to a good conclusion. In fact, it might be better to use a wider range of possibilities to broaden our scientific scope and ability to make Tori's parents stay together. Anything was better than the failed spider experiment.

A thought rang to life in my head. "I have a proposal."

Roger and Georgia looked at me curiously.

"I propose that we all study this topic independently. We can use our own expertise to try and make Tori's parents stay together."

Roger and Georgia slowly nodded.

"You see, then we don't have to decide whether or not we should use a strict anthropological lens or not.

Georgia can do that. Roger, you can use physics."

Roger nodded his head. "Yes. I see what you mean. You think we should each do our own independent experiments and then come together to see what the best solution is."

"Exactly!" I yelled. My heart was thumping with excitement. "Then Tori can have three different ways to solve her problem."

"Awesome!" Georgia yelled.

"I'm on board," Roger said.

"Then it's settled. Let's make our deadline this Friday. We all meet at my house and discuss our results. I'll ask my mom if we can get pizza."

Roger and Georgia smiled and nodded their heads in agreement.

"Pepperoni or supreme?" Georgia asked.

I pressed my finger to my lip. "I'll ask for both. And maybe a cookie pizza, too."

Georgia smiled and licked her lips. "Awesome!"

"Then it's settled," I announced. "We come together on Friday and present our findings to Tori."

"I'm reminded of an excerpt from a Robert Frost poem," Georgia said, looking skyward. She pushed her glasses up her nose and closed her eyes.

"Oh brother," Roger muttered out of the corner of his mouth. "Here we go."

I punched him in the arm.

"Ouch!" he yelled as he rubbed his arm.

Georgia began reciting.

CHAPTER FOURTEEN

My mom has always been a beacon of sound advice, and not just because she's my mom and made me feel better when I had Hand, Foot, and Mouth Disease and looked like you could connect the dots on my body. And not just because she helps me with my Spanish homework when I want to give up because I can't stand learning foreign languages. (Sometimes, it gets so bad I swear off going to Jimmy Taco's on Taco Tuesday just so I won't have to say "*hola*" to Señor Perez, my favorite waiter.) I respect Mom's opinion, mostly, because she has done some pretty wickedly amazing things in science herself.

My mom is a plant physiologist who specializes in plant communication. That's right, plant communication. Plants actually send messages back and forth to each other. Not the kind of messages that Apollo and Mandy send back and forth, about how much they love each other and basketball or other boring topics, but messages about the environment, like when a frost is coming or a heat wave. Plants also have been shown to send messages to certain beneficial insects. If a plant is getting hammered by a pack of leaf-munching bugs, for instance, they send signals to beneficial predators like ladybugs to get some help. The beneficial insects then swarm in and eat all the leaf-munching bugs, doing the plants a solid and getting some much needed grub at the same time.

When my baby brother Horton was born, my mom decided to take a hiatus from her full-time scientific exploits. She said Horton needs extra care with his special needs and all. Now she spends lots of time doing her second favorite thing: writing romance novels. Her pen name is Lady Demure and she does pretty well. She hasn't broken the *New York Times* Best Sellers list, but it's only a matter of time.

"Mom," I said. "Can I get some advice from you?"

Mom was feeding Horton green goop from a tiny jar. He was batting the spoon and making the goop splash everywhere. I shuddered.

"Sure, Sera. What's on your mind?" she asked.

Horton grabbed the spoon and threw it across the room. Madame Curie, our wiener-beagle dog, jumped up and caught the spoon. She began to shake it back and forth.

"No, no, Madame Curie," Mom said as she chased the dog.

"No, no," Horton repeated from his chair.

Mom grabbed the spoon in Madame Curie's mouth. The dog wouldn't let go. Mom pulled with all her might and lifted the dog at least three feet in the air.

"You've already had lunch, girl. You don't need to eat a spoon."

The dog wiggled and growled in the air. She looked like a giant, squirming fish.

Horton giggled loudly and clapped his hands.

"Oh, you like that?" Mom asked and smiled at Horton.

Finally, the dog released her grip and dropped to the floor.

"Phew," Mom said, wiping her forehead. "We're down to our last spoon. I think that dog has a taste for plastic."

I giggled as Madame Curie ran in happy circles around Horton's high chair.

"I think the plastic is affecting her brain," I said.

"Yeppers," Mom said. "It must be the plastic." She wiped Horton's face and pulled him out of his seat. "What's on your mind, Sera?"

I collected my thoughts and began. "I'm developing some scientific protocol, and I'm not sure where to begin."

"Okay. Well, what exactly are you studying?"

"I'm not actually studying anything. I'm trying to solve a problem."

Mom kissed Horton on the head and put him in his playpen. "What problem is that?" she asked.

"Well," I said taking a deep breath. "It's the problem with Mr. and Mrs. Copper."

Mom tilted her head in a sympathetic, mom kind of way. "Oh, Sera. No. That's not a problem you can solve."

"Why not?" I yelled, possibly a little too loud.

"Sera," Mom scolded. "Lower your voice."

"Sorry," I said sheepishly. "I think I must've let my passion get out of control."

Mom nodded her head.

"I just need some advice on how I can begin developing a good working hypothesis. I'm at a dead end. I have been reading up on biology and chemistry and even psychology. Psychology, Mom! I'm delving into the liberal arts!"

"Calm down, Sera. Psychology is technically a science," she said in a soothing voice.

"If you say so," I muttered under my breath.

"Serafina. The problem between Tori's parents is not one that can be solved through a scientific experiment."

I stuck out my bottom lip.

Mom tilted her head again, indicating that I was beginning to soften her up. I was getting her exactly where I wanted her.

"What exactly did you have in mind?" she asked.

I thought for a moment. "Obviously something dramatic. Maybe cold fusion or radiation or even re-animation." I thought about the pack of zombie roadkill raccoons I was planning to unleash on Apollo.

Mom chuckled. Her reaction was not exactly what I was expecting.

I began to pout again, this time with full lip and watery eyes.

"Sera, that's not what I meant at all," she said,

pulling me into a hug. She pressed my face against her chest, squishing the left side of my head. "I know you mean well. But I think you're going to have to leave this problem up to the adults."

"What?" I said, pushing away from my mom. "I can't do that to Tori. She's my best friend."

Mom studied me. Her pretty eyes, the color of green sea-glass, flickered across my face. "Okay. I understand."

"You do?" I asked, wiping my nose with my fist.

She nodded her head. "Yes, Serafina. I do understand. And yes, I'll help you develop a theory. But I want to warn you, I'm not sure we can actually solve the problem. The best you can hope for is something that may provide an explanation."

I squinted my eyes at her in confusion. "Just an explanation?"

"Sure," she said. "You see, science doesn't always just solve problems. Science also provides explanations."

"What do you mean?" I asked.

"Well, like cellular division: meiosis and mitosis. There's not necessarily a scientific problem associated with cell division. Meiosis and mitosis were originally discovered and described so we could have a better understanding of how cells reproduce."

I nodded my head. "I think I see where you're going with this. Meiosis and mitosis don't solve a problem, they just explain what's happening."

"Exactly," Mom said, pulling me in for another hug. "So the most you can hope for is discovering an explanation as to why this is happening. Not necessarily solving the problem."

I hugged Mom and breathed in her scent. She smelled like vanilla cookies and that green goop she had been feeding Horton. "I understand, Mom. Thank you."

"Of course, baby. I love you." And she kissed me on the head.

CHAPTER FIFTEEN

Until then, the worst thing that had ever happened to me was when a truck hit my dog, Nikola Tesla. Nikola was a lab mix and loved to dig holes in our backyard. Then he would wiggle under the fence and escape. Our neighbors have a small picket fence and Nikola often would hop right over it and run around the neighborhood for hours. He also dug up small trees and tried to make random people play fetch with him. Sometimes he rang people's doorbells with his wet nose.

One time, he brought a six-foot-tall pear tree to Mrs. Gomez, our backyard neighbor. Nikola rang Mrs. Gomez's doorbell twice and when she opened it, he ran right past her and put the tree on her couch and tried to bury it under the cushions. Mom had to vacuum mulch from Mrs. Gomez's house for almost two weeks after the incident.

One day, when Apollo and I were getting off the school bus, both Mom and Dad were waiting for us on the front porch. Dad was holding Nikola's collar. Mom's eyes were all red and swollen from crying, and her voice was shaky.

Apollo took one look at my parents and got it. He knew something bad had happened to Nikola. It wasn't so easy for me. It was like my brain couldn't accept the idea that something bad could've happened to my dog. Dad had to tell me about the pickup truck and Nikola running out in the street. After that, I couldn't concentrate for a week. Everything felt bad, kind of sideways in my life. It was the same way I was feeling about Tori and her parents' situation. I was all mixed up inside.

Mom and I spent a lot of time researching. Sometimes our eyes felt red and stingy, but we kept on reading books and surfing the net. As I put in the long hours working on Tori's problem, what kept me going was the thought of what a great friend Tori was to me.

Tori and I did almost everything together. We collected bugs, watched the stars, watched movies together, shared food and friends and secrets. She was like my sister. I couldn't let her down. But her problem—the problem of her parents' divorce—seemed to have no solution. It looked like, for the first time ever, science was letting me down.

I kept thinking about all the super harsh things people had said to me over the past week, like the stuff about how it was a "grown-up" problem. When people say things to me like certain issues are only meant for "grown-ups" or "mature individuals", it makes me super-crazy mad. The reason I get so flaming hot is because people sometimes put things into categories that are unreachable to kids on purpose. And it's not because kids aren't capable of coping in our own way. I've spent the past two summers experimenting with alternative propulsion systems for space exploration. I think I can handle it.

I've always thought that inherent in any problem is a solution. Any person—adult or child—with persistence and a willingness to solve the issue can find the answer, one way or another. And that is exactly what I was going to do; I was committed to finding the answer for my best friend, Tori. And at nine on Thursday night, after Dad and Mom and Horton and even Apollo had gone to bed, I found the answer I'd been looking for all along.

Friday night came faster than I expected. At 6 p.m., the doorbell started ringing. First it was Georgia, then Roger, then the pizza, and then Tori.

Mom had stayed true to her word and ordered a supreme pizza, a pepperoni, and a cookie pizza. She even ordered a mushroom and olive for Apollo and Dad and anchovy with jalapenos for herself. Dad almost barfed when he saw her pizza! As my friends and I ate, we joked and laughed and seemed to forget about the serious issue that had been plaguing us all week.

Roger told us about a miniature pinscher he was walking every other day and how the dog had unexpectedly jumped up and bitten Roger's rear end. The bite was so hard it made a hole in Roger's jeans! Georgia told us about her dad's new hobby: competitive eating. She said her dad had to practice at every meal so her mom had made 100 hot dogs with chili, cheese, and mustard for dinner. Her dad ate over 80 dogs in under three minutes! With his hands tied behind his back!

Tori roared with laughter and seemed to be brighter and more carefree than she had been all week. Her mom was planning to move out at the end of the month and she wasn't taking Tori. She was going to move back to Germany so she could take a job there. Tori and her dad were going to stay in Texas. Tori had calculated the exact mileage from her house to the city where her mother was going to be living. Let's just say the number was large. Large enough that she started to cry again when she saw all those digits on her calculator.

Each time I had spoken to Roger or Georgia, neither one seemed closer to solving the divorce problem. I could tell they were as troubled as I was. But if I knew my friends as well as I thought I did, they would have something that would help Tori and her parents.

"Did you hear they found Doug today at lunch?" Roger asked.

"No," I said. "Finally!"

Roger nodded his head. "He was in Todd Brakefield's Lunchable. Todd was layering his ham and cheese when he felt something furry."

"No way!" Tori screamed with laughter.

We all roared with glee imaging Todd touching one of Doug's furry legs.

At 7:30 p.m., we went up to my bedroom to

commence our meeting.

"Alright, everyone," I said, gesturing to my friends. "We've gathered here tonight for a very important reason."

Heads nodded in agreement.

"Tori, our good friend and trusted colleague, is undergoing a very serious issue in her life." Tori bowed her head slightly. I continued. "But we are all her friends and have been working very hard on her behalf."

Georgia and Roger murmured in agreement. Something seemed unsteady and unsure in their voices. I looked both of them in the face. Roger looked away. Georgia held my gaze as she enthusiastically hugged her Lisa Frank notepad.

"As you all know, this problem has not been an easy one," I said as Tori looked at us. "Divorce seems to be something that science has yet to unravel."

Tori blinked her eyes. "So does that mean you all have bad news for me?" she asked, her voice nothing more than a whisper.

"No," Georgia said, pushing her glasses up her nose. "Not at all. Do you mind, Serafina?" She motioned toward the rainbow leopard on the cover of her notebook.

I shook my head. "Please," I said and took a seat next to Tori.

Georgia cleared her throat theatrically and tossed a few blonde curls over her shoulder. She cracked opened her notebook, which was covered in writing in all different colors. It looked like she had been working hard to solve Tori's problem.

"As you know, I approached this issue from a person-centered view. I believe that since this is an issue that affects people like yourself, theories about human behavior and culture can provide the most insight."

Normally statements like that from Georgia would elicit a litany of moans and groans from Roger. Roger is a strict materialist. He believes that everything in the natural world can be explained by matter and substances and chemistry and reactions (basically anything and everything that can be observed). Roger can't stand feelings or behavior or anything that he can't stick under a microscope. But, all he did was sit there; he seemed deflated, defeated. I gave him a curious glance. He kept his eyes trained on his bright red high tops as Georgia continued.

"I think we must first define marriage to get a better understanding of the nature of divorce. Western civilizations have often viewed marriage as a social or religious construct. Marriage was, and is, viewed as the foundation of the family."

At this, Tori grimaced. She was probably thinking about how her own family was crumbling.

Georgia turned the page to another one filled with more colorful scribble. "Very often, the nature of marriage and how it is defined is dictated by the present social and cultural conditions. Today, people get married because they fall in love. A few hundred years back, your parents decided who you would marry. A few hundred years from now, it will likely be different again." She turned another page. "Today, when a marriage ends in divorce, the parents live apart and they choose where the children live. Much like what is happening to you, Tori." Georgia took a deep breath. "And this leads me to my theory. Tori, based upon the changing nature of how marriage has worked and has been defined in historical context"— she swallowed—"it's very possible that in the near future a marriage that ends in divorce could actually mean that both parents have to take their kids and live at Disney World for the rest of their lives instead of splitting up." Georgia slammed her notebook shut. "The end," she said and pressed her glasses back up her nose.

"Oh, come on!" Roger yelled. "Is that all you have, Georgia? That *maybe*, in a few hundred years or so, divorce might be nice and won't be so bad because parents would be forced to stay together and ride Space Mountain until they puke?"

Georgia scowled at Roger. "I never said anything about Space Mountain."

Tori looked flushed.

"Okay, okay," I said as I stood between Roger and Georgia. "Let's not fight."

"Look," Georgia said with great exasperation. "That's the best I could do. Everything I read about divorce was pretty straightforward. Divorce happens when two people can't get along. It's been happening for centuries. Sometimes two people just weren't meant for one another." She put her hand on Tori's. "I'm sorry, Tori. I wanted to do better."

Tori shook her head. "I know, Georgia. You're a great friend."

Georgia spun around to Roger. "Let's see you do better," she hissed at him.

Roger straightened his back and his eyes grew wide. "Well," he said as he rose to his feet. "I do have some working theories ..." His voice trailed off.

"Oh really?" Georgia asked. "Well, why don't you just *wow* us with your brilliant insights, Poodle Poop?"

Roger pointed his finger at Georgia. "That was a low blow. And since you mentioned it, I think I prefer Professor Poodle Poop. It has a nice ring to it."

Georgia rolled her eyes while Tori snickered.

"So," Roger began. "I'm sure you're all familiar with Newton's three laws."

"Oh, brother," Georgia wailed.

"Just wait a second," Roger said. "The third law states that for every action there will be an equal and opposite reaction. So what we might have here is a simple case of reaction gone too far."

"How do you figure?" Tori asked.

"Well, obviously this whole divorce thing has been caused by some action in the past that is creating the motion or the action here in the present. What we need to do is to identify what the source of the action is and cut it off, right where it begins." He made a chopping motion with his hands.

"Boo!" Georgia shouted from the bed, her hands cupped around her mouth. "The Professor has something stinky on his theory."

Roger's mouth pinched into a tight line while all three of us girls laughed.

"Come on," he moaned. "This was a really hard problem. I worked day and night thinking about this thing. I can't come up with any viable solution. Tori, I'm really sorry," he said as he fell onto the floor. He splayed his arms and legs wide open in defeat.

"It's okay, Roger. I understand." Tori held her hand out to help Roger back up.

He covered his face with his hands. "I failed," he said through his fingers.

"Get up, Professor Poop," Georgia said and nudged him with her foot. "This was a hard problem."

Roger rose and hugged Tori. "I really am sorry, Tori. I want to help, but I feel like there's nothing I can do."

I rose to my feet slowly. "You guys haven't heard what I found out yet." My knees felt weak and wiggly. My heart was beating pretty fast.

"Serafina, there's no way you figured this out. This problem is impossible," Tori said.

"No, I did," I announced on shaky breath. "I had a breakthrough last night."

Tori, Georgia, and Roger exchanged sideways glances.

I pulled the paper I had taken my notes on from my pocket. The letters swam across the page. I couldn't seem to focus. I was too nervous. I crumbled the paper in my hand, tossed it on the floor, and began.

"Tori. First of all, I wanted to say that I didn't actually solve your problem."

Tori groaned. "Sera, you got my hopes up. Just sit down. I'll be fine. Parents get divorced all the time. It's just something I'm going to have to deal with," she said with a defeated tone.

I shook my head. "No," I said. "You don't understand. I admit I couldn't *solve* your problem. But I did something better. I found an explanation that I think will help."

My friends looked at each other again in confusion.

"Go on then," Georgia said.

I straightened my back and closed my eyes. "You see, Tori. It has to do with quantum entanglement."

Roger muttered, "Interesting," and curled his lips into a smile.

"Quantum entanglement is a principle in quantum physics. I know physics is usually Roger's bag, but this seemed like the perfect theory to create understanding around your parents' divorce." Tori nodded her head. "I know that one of the hardest things about the divorce is the fact that you and your mom will be separated by thousands of miles. She'll be living in a totally different country and you won't be able to see her every day."

"That's true," Tori agreed.

My stomach began to settle down and my heart started to slow. I began to feel normal again. I took another deep breath. "You see in the principle of quantum entanglement, once two quantum elements like an electron or neutron have been observed together, they can no longer be described as independent objects."

"I'm not following," Georgia said. "Can you provide some translation on that?"

I nodded my head quickly and placed my hands on my stomach. "Okay, sure. You see, an atom is made of three different types of particles: protons, neutrons, and electrons. They all have a specific role in the operation of an atom."

Roger added, "Yeah. Each particle has its own charge. Protons are positive, electrons are negative, and neutrons have no charge."

"That's right," I said. "And their charge determines where they exist in the atom. Each particle has a specific responsibility. Kinda like in families. Moms and dads have specific roles and so do kids."

"Okay," Tori said with growing interest.

"Well, not all atoms stay together. Sometimes they combine with other atoms to form different materials and the particles are displaced."

"What do you mean 'displaced'?" Tori asked.

"Well, I mean that an electron that was once with a proton may be put onto an entirely different atom and have to exist and work with different protons."

"Okay," Georgia said. "So atoms may change their particles like a family may have to change when a divorce happens?"

"That's right," I confirmed. "But here's the thing. The principle of quantum entanglement states that once subatomic particles have existed together at the quantum level, like they have formed an atom together, they can no longer be considered independent of each other."

"That makes no sense!" Georgia exclaimed. "I mean, if they're not working together on the same atom, what happens if they are moved like a hundred miles away. How can they still be considered together?"

"That's the thing!" I shouted loudly. "That's the great part! You see, when the particles are together on an atom, they move in specific ways based upon their role. They have to work together. Electrons move in a spin-up and spin-down way, meaning when one electron is at the top of the atomic orbit, the other electron is at the bottom. They act that way to provide stability for the atom. But here's what scientists found out: If you break up those electrons, they continue to move in the same spin-up and spin-down pattern they've always used. Even if you were to move them across the universe!"

"So, even if 5,307 miles separate them, it's like they're still together because they always function the same way?" Tori asked.

"Exactly," I shouted. "You see, Tori? You and your mom are very close. I mean, she's your mom. You guys have been together since you were a little baby. You can't take that away. You guys have been entangled. So at the quantum level, even if she moves to a totally different country, you'll always be together."

"I'm not sure if I get that," Georgia said.

"It's like this," I said. "Just like the electrons that once existed in the same orbit and share the spin-up and spin-down pattern, Tori and her mom are in sync. Every time Tori's heart beats so does her mom's. Even if Tori can't see her mom every day, each time she feels her heart beat she'll know her mom's heart is doing the same. They can never be viewed as completely independent. They've shared a home and life for 11 years now. Even if 5,000 miles separate them, they'll always be a family."

"Wow," Roger said. "That's heavy."

I smiled broadly and nodded my head. "You see, Tori? The divorce is going to happen no matter what Roger or Georgia or I have tried to discover through the scientific method. We can't change the fact that your parents have made a decision that will affect you." Tori nodded her head. "But we are scientific enthusiasts and that gives us a better perspective than most."

"Totally," Georgia said smiling.

"So, we couldn't make your parents stay together, and that was completely a lame-o bummer. And for that I'm sorry. But hopefully, we shed some much needed light onto the dark parts."

Tori smiled. "You did, Serafina." She hugged me tightly.

For the first time in nearly a week, I felt pretty good about life. I certainly wasn't confused anymore.

"You guys are great friends. Thank you all for being there for me," Tori said.

"We're always here for you, even if we can't figure it out," Roger said.

"You did a great job, Professor Poodle Poop," Tori said, giggling.

Roger's face turned as red as his Chuck Taylor

high tops.

CHAPTER SIXTEEN

Science is my jam. Most days. I love science because it is the one thing I can turn to in basically any situation. If Apollo and his gross girlfriend Mandy are all sucky-face, kissy-barfy, then BOOM! Science! I devise a plan to create an army of re-animated road kill raccoons to bite their disgusting, kissy faces off. If my baby brother Horton's smelly diapers are getting my mom down and stinking up the upstairs, then BOOM! I invent an automatic diaper-crushing machine that efficiently and rapidly smashes down the baby doody and smell. Sure, I had a slight miscalculation regarding the doody-bounce back factor after being met with several hundred pounds of pressure, but I worked through those obstacles.

All of this is guided by the scientific method. It's the foundation by which all science is cultivated. It's like the skeleton in a body, the nucleus of a cell. It's the inner core that drives an entire discipline of human understanding. Therefore, in theory, the scientific method should never fail. Right?

When Tori told me her parents were getting a divorce, I knew it was a big deal. Not just because I had always heard that divorces are totally a big deal, but because of the way it made Tori feel. The whole divorce thing was way harsh on Tori. Her brain became distracted, her eyes became watery, and she lost interest in all her favorite hobbies, like bug collecting and meteor watching. Tori turned into a person I didn't know.

It was no mystery then when I turned to science to help solve Tori's parental problem. I stuck to what I knew: the scientific method. I knew I wasn't a marriage counselor or a miracle worker, but I was a kid who loved science and her best friend.

Step one was a no-brainer. My purpose was to make Tori Copper a happy person again. I wanted to make sure she could enjoy bug hunting, and star watching, and pizza eating once more. The only way I knew how to make my best friend happy again was to make sure her parents were happy.

Step two was a little trickier because I wasn't sure about the whole divorce thing. What I knew about it involved some pretty simple calculations. Basically, I knew that divorce was the opposite of marriage. And if marriage was all about love and kissing and bliss, then I needed to make that happen for Tori's folks.

Because Tori was my best friend, I knew some things about her parents, like what they did for a living and what they usually cooked for dinner on Thursday nights (roast chicken and potatoes). What I didn't know was how to make them fall back in love.

I've watched enough romantic movies with my mom to know what makes people fall in love is lots of extraordinary circumstances. Mr. Copper kind of looked like the male lead in a movie, so I developed my hypothesis, my step three. I theorized that if I could create a situation where Tori's parents were forced to deal with an extraordinary crisis, they would fall back in love again. It would be exactly like the way people fall in love in all of my mom's sappy movies.

Step four meant that it was experiment time. Mr. Hairy Arms' collection of spiders happened to be just the thing I needed to create extraordinary circumstances for a romantic reconciliation. My experiment involved creating a situation where Mr. Copper and Mrs. Copper would be forced to come to the school, at the same time, to deal with a crazy situation. I believed this experiment would force them to put their differences aside and fall back in love. And the Coppers falling back in love was exactly what I needed to make sure my best friend would be happy again.

Of course, that wasn't the way that it happened at all. In fact, during my step five, my analysis, I considered that the only thing I actually created was more trouble for Tori. There was no reconciliation, no kissy, sucky-face. There was only scolding from the adults and punishment for my friends and me, including Tori.

I understand now that I went about my experiment in the wrong way. My original purpose was to make my best friend happy. I believed that by making her parents happy, that I could make Tori happy. What I didn't know was that people, especially adults, cannot be controlled, even by science. Sometimes, the reality is that two people—even if they are your best friend's parents—are not meant to be married.

Instead of me focusing on getting Tori's parents back together, I should have focused on helping my best friend focus on her own happiness. Happiness comes from lots of different things: Disneyland, pizza buffets, petting a tarantula's head, and the people around you. Happiness can also come from reconfiguring your idea of what a family really is all about. Sure Tori's parents were not going to be together anymore, but that didn't mean that Tori should be sad all the time.

In the end, Tori's folks did get a divorce and her mom did move to Germany. Tori no longer got to see her mom when she left for school in the mornings or in the afternoons when she got off the bus. She didn't get to spend Friday nights in her mom's bed, watching chick flicks and eating mint chocolate chip ice cream. However, something stirred from within. When she felt lonely and sad and thought she couldn't go on anymore because she missed her mom so much, her heart would beat. One strong thump. Sometimes it was as loud and hard as a snare drum.

Each time that happened, Tori knew—at that exact moment, 5,307 miles away—her mom's heart was doing the same thing, and she didn't feel so alone. No matter the time or space or divorce papers, Tori and her mom were always family. They were forever entangled.

FROM THE WEBLOG OF SERAFINA STERLING

www.SerafinaLovesScience.com

Serafina's Scientific Stylings

Entry #2—The Stink of Failure

I remember my dad telling me once that science is about trial and error with a heavy emphasis on the error. He told me that scientists often learn more from their mistakes and failures than they ever do from their successes. I told him that he sounded like an incompetent professional who was clearly not doing his job correctly. He told me I was not allowed to speak to him like that and sent me to my room.

His words stuck with me, though, and not just because I pouted the rest of the afternoon in my room while everyone else got to eat popsicles on the patio, but also because what he was saying was so different from what I thought a real scientist was. I have always believed that a scientist should strive for perfection. I believed that when you let error or miscalculations or total failure creep in, you had failed as a professional, as an individual. I thought these mistakes said something about the person you were on the inside.

While I have always endeavored for perfection, I have certainly made my share of mistakes. There have been the unfortunate explosions, noxious chemical clouds, escaped class pets. No matter what I had done in my career as a scientist, though, I had never let one of my friends down. I especially had never let someone down when they were in a time of need.

When my grand experiment to reunite Tori's parents failed, I thought that I had failed as a person. I believed that I was not worthy as a scientist and also as a friend. I felt helpless and unable to do anything to help my very best friend. What I didn't see was that it was a situation that was not going to be fixed. Nothing about her parents' relationship was ever going to be normal again. Tori's new normal was the fact that her parents were no longer going to be married. They were going to live in separate towns, separate countries, and she was not going to live life with both of them at the same time. Tori was going to have to live life with her parents in a new way, a divorced way.

What I failed to realize is that science hadn't failed me; I simply had been looking at the problem in an incorrect way. What I should have realized is that the principle of quantum entanglement was there all along, with its subatomic spookiness and theoretical simplicity. Quantum entanglement did not "solve" Tori's problem for her, but it did help redefine and reframe the issue. Yes, Tori would have to accept that her parents would never be in love again, but this did not negate her love for them. Tori's love for her parents was the quantum element that would entangle their hearts forever.

I told my dad today that I understood what he meant about learning more from your mistakes than your successes. I told him how complicated and multi-layered problems can be and that failing on a first attempt very often leads to a better pathway to "solving" a problem. He smiled, and my heart melted. I really thought we were going to have a father-daughter connection at that moment. Instead, he stuck his tongue out and said, "I told you so."

THE DOUBLE SLIT EXPERIMENT

As my friend Tori demonstrated, life can be very complicated. Things are not always as they seem. As a scientific enthusiast, I know this is a truth that comes from the most fundamental of places: the quantum level.

For lots of years, scientists were duking it out over one simple issue: is light a wave or a particle? Each wave-r and particle-r had his or her own ideas and evidence to back it up. So much so that it seemed like the issue would never be solved. (I can only imagine all the wig pulling and glove slapping that must have gone on in those early days.) Then in 1801, Thomas Young came along and created an experiment to end the debate.

The double slit experiment is now a classic experiment in the physics world that proves something that might seem totally impossible at first. Light is neither a wave nor a particle; it's both. That's right, both!

The idea is that light exists in a dual reality. It can show up as a wave or as a particle, depending on the situation and the observer present. Try to process that for a moment. What I am saying is that light can be either a wave or a particle, and it mainly depends on you and your beliefs.

Young's experiment was later expanded to prove that it wasn't only light that existed in a dual state, but also electrons themselves. This principle is called quantum superposition.

Just think about this, the most fundamental particles of our universe are kind of magical. No, scratch that. They are a *lot* magical. They can conform to a reality that is determined by a person. What do you think that says about you and your life? Well, don't take my word for it. Prove it to yourself with this experiment.

What You Need

- One laser pointer—nothing fancy, even one of those cat toys will work
  - Note: Never point this at anyone's eye or at your own. Lasers can seriously damage your eyesight and make your mom very angry.
- One fine-tooth comb—ask your dad or even your grandpa
- Electrical tape
- Blank note card (5x7)
- Two large binder clips (I bet your parents have some in their office)
- Two medium binder clips

Step One

Use the electrical tape to cover the teeth on the comb

leaving only **two** adjacent teeth exposed.

Step Two

Clip one side of the comb with one of the large binder clips, allowing the comb to stand upright on a table or flat surface. Be sure that the two exposed teeth are **vertical.**

Step Three

Using the two medium binder clips, clip either side of the laser pointer. What you are trying to achieve is a holder for the pointer that will allow it to rest horizontally on the table.

Step Four

Use the remaining large binder clip and create a stand for the index card by clipping it to the side of the card. This should allow the card to stand upright on your flat surface.

Step Five

Begin to assemble your experiment by placing your standing index card on a flat surface. Set your laser pointer about four feet away directly facing the card. Last, place your comb with the exposed teeth approximately a half inch from the laser pointer.

Step Six

Turn off the light and fire up your laser! You should see a light pattern on your index card. The pattern should appear banded. This is because your comb, the double slit, is interfering with the light. You will probably see alternating light and dark patterns. If light simply existed as a particle, it would pass right through the two slits with no interference. But, since the world is so weird and magical, you are probably witnessing the wave pattern on your index card right now.

Try covering one slit. See what happens to the pattern on your card. The light is probably behaving more like a particle now and looking more like a concentrated point of light.

Go further:

Ask yourself, why does this happen? More importantly, what does this say about the nature of our reality?

Sources:

https://www.exploratorium.edu/snacks/two-slit-experiment

http://farside.ph.utexas.edu/teaching/302l/lectures/node151.ht

About Cara Bartek, Ph.D.

I love science! More importantly, I love helping kids develop a passion for science. This world is a big, scary, and confusing place sometimes, but the good news is we can rely on some seriously important things to help us through, like faith and family and friends and education! Education helps open the door to opportunities and worlds we may not otherwise experience. It actually makes the world a bigger place! But we have to feed our education, and we do that by scratching our curiosity. Curiosity is the key that helps our brains and our hearts grow. I created the Serafina Loves Science! series to show that science is not only interesting but also relatable. I hope to make the world bigger and brighter and way more interesting through science!

I live in Texas with my husband, Matt, and my two little girls, Caroline and Penelope. I also am compelled to mention my two furry children, Beetle and Bob. Bob is a weiner-beagle (think overstuffed sausage with long legs,) and Beetle is a possible pit/raccoon mix; genetic tests are still pending. We also have a couple of hissing cockroaches named Shimmer and Shine that live in a pink terrarium on our kitchen island. Matt and I own an agricultural business that keeps us busy and sunburnt! Some of my favorite things to do are charting stars with my kids, spending time on the beach checking out the sea turtles and pelicans, reading (my favorite book is *A Wrinkle in Time*), and writing. The rest is history! Visit me at www.carabartek.com.

Did you enjoy this book?

Please consider leaving a review for *Quantum Quagmire* on any online platform, and look for *Cosmic Conundrum*, book one in the Serafina Loves Science! series, as well as upcoming books.

Visit www.absolutelovepublishing.com for the latest releases. And be sure to leave your review on any online platform!

About Absolute Love Publishing

Absolute Love Publishing is an independent book publisher devoted to creating and publishing books that promote goodness in the world.

www.AbsoluteLovePublishing.com

Young Adult & Children's Books by Absolute Love Publishing

Dear One, Be Kind by Jennifer Farnham
This beautiful children's book takes young children on a journey of harmony and empathy. Using rhyme and age-appropriate language and imagery, *Dear One, Be Kind* illustrates how children can embrace feelings of kindness and love for everyone they meet, even when others are seemingly hurtful. By revealing the unseen message behind common childhood experiences, the concept of empathy is introduced, along with a gentle knowledge of our interconnectedness and the belief that, through kindness, children have the power to change their world. Magically illustrated with a soothing and positive message, this book is a joy for children and parents alike!

Different
Twelve-year-old Izzy wants to be like everyone else, but she has a secret. She isn't weird or angry, like some of the kids at school think. Izzy has Tourette syndrome. Hiding outbursts and tics from her classmates is hard enough, but when a new girl arrives, Izzy's fear of losing her best friend makes Izzy's symptoms worse. And when she sees her crush act suspiciously, runaway thoughts take root inside of her. As the pressure builds and her world threatens to spin out of control, Izzy must face her fear and reveal her secret, whatever the costs.

Authentic and perceptive, *Different* shines a light on the delicate line of a child's hopes and fears and inspires us all to believe that perhaps we are not so different after all.

The Adima Chronicles by Steve Schatz

Adima Rising
For millennia, the evil Kroledutz have fed on the essence of humans and clashed in secret with the Adima, the light weavers of the universe. Now, with the balance of power shifting toward darkness, time is running out. Guided by a timeless Native American spirit, four teenagers from a small New Mexico town discover they have one month to awaken their inner power and save the world.

Rory, Tima, Billy, and James must solve four ancient challenges by the next full moon to awaken a mystical portal and become Adima. If they fail, the last threads of light will dissolve, and the universe will be lost forever. Can they put aside their fears and discover their true natures before it's too late?

Adima Returning
The Sacred Cliff is crumbling and with it the Adima way of life! Weakened by the absence of their beloved friend James, Rory, Tima, and Billy must battle time and unseen forces to unite the greatest powers of all dimensions in one goal. They must move the Sacred Cliff before it traps all Adima on Earth—and apart from the primal energy of the Spheres—forever!

Aided by a surprising and timeless maiden, the three light-weaving teens travel across the planes of existence to gain help from the magical creatures who guard the Adima's most powerful objects, the Olohos. There is only one path to success: convince the guardians to help. Fail and the Cliff dissolves, destroying the once-eternal Spheres and the interdimensional light weavers known as Adima.

Like the exciting adventures of *Adima Rising*, the second spellbinding book of The Adima Chronicles, *Adima Returning*, will have your senses reeling right up until its across-worlds climax. Will conscious creation and the bonds of friendship be enough to fight off destructive forces and save the world once again?

The Soul Sight Mysteries by Janet McLaughlin

Haunted Echo
Sun, fun, toes in the sand, and daydreams about her boyfriend back home. That's what teen psychic Zoey Christopher expects for her spring break on an exotic island. But from the moment she steps foot onto her best friend Becca's property, Zoey realizes the island has other plans: chilling drum beats, a shadowy ghost, and a mysterious voodoo doll.

Zoey has always seen visions of the future, but when she arrives at St. Anthony's Island to vacation among the jet set, she has her first encounter with a bona fide ghost. Forced to uncover the secret behind the girl's untimely death, Zoey quickly realizes that trying to solve the case will thrust her into mortal danger—and into the arms of a budding crush. Can Zoey put the tormented spirit's soul to rest without her own wild emotions haunting her?

Fireworks
Dreams aren't real. Psychic teen Zoey Christopher knows

the difference between dreams and visions better than anyone, but ever since she and her best friend returned from spring vacation, Zoey's dreams have been warning her that Becca is in danger. But a dream isn't a vision—right?

Besides, Zoey has other things to worry about, like the new, cute boy in school. Dan obviously has something to hide, and he won't leave Zoey alone—even when it causes major problems with Josh, Zoey's boyfriend. Is it possible he knows her secret?

Then, one night, Becca doesn't answer any of Zoey's texts or calls. She doesn't answer the next morning either. When Zoey's worst fears come true, her only choice is to turn to Dan, whom she discovers has a gift different from her own but just as powerful. Is it fate? Will using their gifts together help them save Becca, or will the darkness win?

Discover what's real and what's just a dream in *Fireworks*, book two of the Soul Sight Mysteries!

Serafina Loves Science! by Cara Bartek, Ph.D.

Cosmic Conundrum
In *Cosmic Conundrum*, sixth grader Serafina Sterling finds herself accepted into the Ivy League of space adventures for commercial astronauts, where she'll study with Jeronimo Musgrave, a famous and flamboyant scientist who brought jet-engine minivans to the suburbs. Unfortunately, Serafina also meets Ida Hammer, a 12-year-old superstar of science who has her own theorem, a Nobel-Prize-winning mother, impeccable

fashion sense—*and* a million social media followers. Basically, she's everything Serafina's not. Or so Serafina thinks.

Even in an anti-gravity chamber, Serafina realizes surviving junior astronaut training will take more than just a thorough understanding of Newton's Laws. She'll have to conquer her fear of public speaking, stick to the rules, and overcome the antics of Ida. How will Serafina survive this cosmic conundrum?

Quantum Quagmire
Serafina suspects something is wrong when her best friend, Tori Copper, loses interest in their most cherished hobbies: bug hunting and pizza nights. When she learns Tori's parents are getting a divorce and that Tori's mom is moving away, Serafina vows to discover a scientific solution to a very personal problem so that Tori can be happy again. But will the scientific method, a clever plan, and a small army of arachnids be enough to reunite Tori's parents? When the situation goes haywire, Serafina realizes she has overlooked the smallest, most quantum of details. Will love be the one challenge science can't solve?

Join Serafina in another endearing adventure in book two of the Serafina Loves Science! series.

Also, be sure to check out our
Adult Fiction and Non-Fiction Books!

**Learn more about our books and upcoming releases
at AbsoluteLovePublishing.com.**

Made in the USA
Columbia, SC
12 December 2018